How to Think Like a Neandertal

How to Think Like a Neandertal

THOMAS WYNN
AND
FREDERICK L. COOLIDGE

OXFORD
UNIVERSITY PRESS

OXFORD
UNIVERSITY PRESS

Oxford University Press is a department of the University of Oxford.
It furthers the University's objective of excellence in research, scholarship,
and education by publishing worldwide.

Oxford   New York
Auckland   Cape Town   Dar es Salaam   Hong Kong   Karachi
Kuala Lumpur   Madrid   Melbourne   Mexico City   Nairobi
New Delhi   Shanghai   Taipei   Toronto

With offices in
Argentina   Austria   Brazil   Chile   Czech Republic   France   Greece
Guatemala   Hungary   Italy   Japan   Poland   Portugal   Singapore
South Korea   Switzerland   Thailand   Turkey   Ukraine   Vietnam

Oxford is a registered trade mark of Oxford University Press
in the UK and certain other countries.

Published in the United States of America by
Oxford University Press
198 Madison Avenue, New York, NY 10016

Library of Congress Cataloging-in-Publication Data
Wynn, Thomas.
How to think like a Neandertal / by Thomas Wynn and Frederick L. Coolidge.
p. cm.
Includes bibliographical references and index.
ISBN 978-0-19-932922-9 (pbk.)
1. Neanderthals.   2. Social archaeology.   3. Ethnopsychology.   4. Cognition and culture.
5. Human evolution.   I. Coolidge, Frederick L. (Frederick Lawrence), 1948–   II. Title.
GN285.W96 2013
569.9'86—dc23
2013011446

The cartoon on page 134 is used with permission of The New Yorker. Copyright © Mike Stevens.
The New Yorker Collection/www.cartoonbank.com.

1 3 5 7 9 8 6 4 2

Printed in the United States of America
on acid-free paper

To our brothers, Tim Wynn and Tom Coolidge,
who inspired, encouraged, and tolerated us our entire lives.

Preface

THIS IS NOT a "how to" book for adolescent boys (of any age), nor is it a manual for survivalists who fear an imminent apocalypse. And it is not a satirical account of any political party, individual, or ideology, however deserving. It is a serious attempt to describe the thinking of a vanished and often maligned population of prehistoric people, the Neandertals.

Neandertals hold the distinction of being the only fossil human to have become a part of the standard vocabulary. Everyone who has had exposure to modern media has heard or seen the term, and most people carry some idea of who Neandertals were, though more often than not it is a caricature drawn from modern cinema or television: *Iceman, Encino Man, Caveman, Quest for Fire* (our favorite). There have been many fine academic books on the subject of Neandertals, some of which have been written for a general audience. They are thorough, and often a bit dull. Old bones and stone tools quickly lose their charm for the nonspecialist. What is often missing from serious accounts of Neandertals is any discussion of the questions many people find most interesting: What was it like to be a Neandertal? How were their lives similar to or different from ours?

This is the task we have taken on: to provide an account of the mental life of Neandertals, to the extent that it can be reconstructed from the fossil and archaeological remains. We are an archaeologist (TW) and a psychologist (FLC) who have been working together for a decade on topics in the evolution of cognition. Our approach uses established concepts in cognitive science to interpret paleoanthropological remains. We will describe specific archaeological remains, use explicit descriptions of cognitive abilities, and occasionally ask readers to accompany us on thought experiments. As we hope to show, it is in fact possible to describe some of the ways Neandertals thought about their world. In the process of painting this picture we will present some interesting but little appreciated insights into cognition in general. In the end, our account of Neandertal cognition should provide a perspective from which to view our own cognition, much of which we share with Neandertals, but some of which we do not.

Contents

How to Think Like a Neandertal

True Grit

Who Were the Neandertals?

Neandertals were prehistoric people who evolved in Europe and flourished between about 200,000 and 30,000 years ago. Right off the bat, with this seemingly innocuous sentence, we run into controversy. Referring to Neandertals as "people" is in fact a bit risky. Some in anthropology and psychology would prefer to reserve the term "people," and especially the term "human," for individuals and groups indistinguishable in all respects from people living today, at least biologically indistinguishable. They would prefer that we use the term "hominin" or "hominid," which encompass living people and our ancestors and a number of bipedal cousins, going all the way back to our evolutionary split from apes some eight million years ago. They have a point: the terms "human" and "people" are certainly loaded with implications about behavior and cognition. This is precisely why we choose to use them. Neandertals were so similar to us anatomically and genetically that we believe the default position—the null hypothesis, for readers familiar with scientific method—should be that Neandertals were no different. We must then present sound arguments for any differences we propose. Our initial sentence contains a second controversy: "Neandertal" or "Neanderthal"? My word processor, for example, prefers the latter. When the first Neandertal was recognized and described in the 1850s, the German word for valley was spelled *thal*, and the find was made in the Neander Valley, hence Neanderthal. In the twentieth century the German spelling for *thal* changed to *tal*. Following an adherence to Linnaean rules for biological naming, some scholars argue that the original spelling has priority. Those who favor separate species status for Neandertals are also likely to take this stance, because the formal species name for Neandertals, if Neandertals were a separate species, is

Homo neanderthalensis, a term introduced in the nineteenth century. Others argue that "Neandertal" is not a formal scientific label, and that we should follow current German usage. Take your pick; the terms are pronounced identically (i.e., no English "th" sound).

Neandertals evolved in Europe from an earlier hominin known as *Homo heidelbergensis*. We "modern" humans also evolved from *Homo heidelbergensis*, but in Africa. Fossil and genetic evidence indicates that these two groups separated from one another about half a million years ago, long enough for differences to evolve, but perhaps not long enough for them to become separate species, which would require that they not be able to breed and produce fertile offspring. At first Neandertals' ancestors were restricted geographically to Europe, but by the time true Neandertals appeared about 200,000 years ago they had begun to move into western Asia. Eventually they moved as far east as Uzbekistan (north of Afghanistan) and throughout much of the Middle East. The last Neandertals appear to have been restricted to the Iberian Peninsula (modern Spain and Portugal), where the last one died sometime after 30,000 years ago. Modern humans, who evolved in Africa, moved into western Asia after 60,000 years ago and into Europe by 40,000 years ago. Most paleoanthropologists believe that modern humans were somehow implicated in the disappearance of the Neandertals, but exactly how this happened is unclear and the source of much controversy in the field.

So what were Neandertals like? If they were as similar to us as the evidence suggests, would it even be possible to recognize one if he or she were dressed in modern clothing, sitting or standing at a bus stop? The answer is yes, but you might have to look closely, risking a socially awkward moment. You would first notice the face. Our transit passenger would have a large face that was especially long from the top of the forehead to the chin. He (for simplicity's sake we will refer to him as a he; the features apply to Neandertal women as well) would also have pronounced brow ridges, which are swellings of bone immediately above the eye orbits, and his nose would be longer and broader than is typical of modern people. Indeed the whole middle part of his face would appear to be pulled forward. He would lack a protruding chin, largely because his teeth would be larger, on average, than ours and the tooth row longer. If he smiled, you would probably notice that his front teeth (his incisors) were heavily worn down. There are several other distinctive features of his face that anthropologists would recognize on his skull, but that would not be obvious on a living individual (e.g., a retromolar space between his third molar and the ascending part of his jaw; and speaking of third molars, his large jaw would have plenty of room for these wisdom teeth). Such features help anthropologists identify fragmentary remains as Neandertals and can be involved in technical arguments in paleoanthropological circles concerning who is related to whom, but they are not particularly relevant to our current theme. One distinctive feature that you might or might not notice on our bus stop Neandertal would be the size and shape of his brain case, the part of the skull that holds the brain. It would be large, probably larger than yours. It would be longer than yours, and it would be widest midway between the base and the top, so that from the rear it would appear round. (Your widest point is much closer to the top.) If you accidentally picked up his hat, it would almost certainly fall down over your eyes. Neandertal brains were about 10 percent larger, on average, than modern brains, and so were their heads. So, on the whole, our Neandertal would have an unusual face and head, but it probably would not be so unusual that you would consider walking on to the next stop.[1–4]

Below the head our Neandertal would not appear at all unusual if he were sitting at the bus stop. But if you followed him to the gym, you would probably notice some impressive differences. He would be fairly short, about the same height as President Sarkozy of France or Jon Stewart of *The Daily Show*. But he would be much more massively built. His torso would be broad and deep, with wide hips and a barrel chest. He would be heavily muscled and would likely resemble the bodybuilders at the gym, or perhaps the competitive weight lifters. But he wouldn't have had to invest nearly the time or reps to get his physique; he was by nature very muscular. The muscles of his neck, shoulders, chest, back, arms, and legs would be well-developed—perhaps not the sculpted look of the competitive bodybuilder, but certainly the mass. His arms and legs would be a bit short for his stature, especially his forearms and lower legs. This body type is ideal for power. If he finally noticed that you were checking him out, and challenged you at the bench press or to arm wrestle, you would be wise to decline. It is unlikely that any but the strongest of modern people could match him (Figure 1-1). As with the head and face, Neandertals had a number of less obvious characteristics in their bones that anthropologists use for identification, but these need not concern us here.

Taken individually, none of the characteristics we have just mentioned would be outside the range of modern human variation. There are modern people with long faces, others with large noses or big heads (literally and metaphorically), and still others who have relatively short, massive bodies (the Bulgarian Olympic weight-lifting team comes to mind). Some of the minutiae of anatomy are unique to Neandertals, but surprisingly few. One of us (TW) has occasionally sent his beginning-level students out to identify Neandertal-like traits possessed by esteemed professors on campus, an exercise

FIGURE 1-1 A modern human made up to resemble a Neandertal. Sebastian Willnow/ Getty images. Used with permission.

that invariably leads to mirth. (And yes, TW and FLC have occasionally appeared on the list.) What set Neandertals apart from us anatomically are not the individual features, but the combination, the package of traits, if you will.[5]

But wait. Our gym example has probably prompted some of you to ask a very important question: If Arnold Schwarzenegger can go to a gym, work out, and develop a well-muscled, powerful physique, how were Neandertals any different? Maybe their daily lives required them to do a lot of heavy physical labor (it did), and this is what we see in their skeletons, not some innate biological difference.

An adult's anatomy and physiology are the result of a complex interplay of genes and the environment of development, and it is rarely the case that one or the other is the sole determining factor. One of us (TW) has identical twin daughters who are far from being identical. Even at birth there were subtle differences in anatomy. Yet they had the same genes and had developed in the same uterus. As it turns out gene expression is a complex phenomenon. In this case, even though the girls shared the same placenta, they did not receive exactly the same nourishment. Moreover genes are not as narrowly deterministic as science once thought. Even the transcription of DNA to RNA, the first step in protein synthesis, is affected by the features of the nuclear environment, including the presence of proteins that have entered the nucleus from outside. The cell environment can actually alter how DNA is transcribed; that is, the external environment can alter gene expression at this very fundamental level, and the effect can then be passed down through multiple cell divisions, a process termed "epigenetic inheritance." Some of these epigenetic factors work in probabilistic ways, so that an identical DNA sequence might come to be expressed differently in separate groups of cells or individuals. Thus it is not a surprise that identical twin girls are in fact different. Even more significant to their adult anatomy, and more pertinent to our Neandertal discussion, were their differing experiences as children. One girl chose soccer, the other ballet. They exercised muscles differently and stressed bones differently, and this difference had a permanent effect on how their bodies grew. So when we describe Neandertal body types as different from that of modern humans, we must acknowledge that this difference was a complex result of different physical environments, different experiences growing up, and differences in genes.[6]

So what about Neandertals? Take their barrel chests. There are people in the world today who have chests shaped very much like the barrel chests of Neandertals. These are people who have lived their entire lives at altitudes above 3,000 meters (10,000 feet). Because of the reduced oxygen pressure at high altitude, these people develop larger lung capacities and larger chests (and even slightly larger hearts). Little, if any, of this is determined by their genes. Instead their individual bodies have changed to adjust to local conditions. Few, if any, Neandertals lived at high altitude, so this specific explanation will not work. But perhaps their lives of heavy exertion led to enlarged chests; heavy exertion also puts stress on an individual's ability to process oxygen. Similarly, Neandertals' large jaws may in part result from use. Many Neandertals have heavily worn incisors, suggesting habitual, heavy use, perhaps as a kind of third hand. Add such use to heavy chewing, and the jaw will become more massive as one grows up. Modern people who do a lot of heavy chewing have larger jaws (and less orthodontia) than those who don't. It is not a genetic difference; it is a difference in their way of life.

So some Neandertal features can also be explained by their way of life. How can we tell which features were innate and which developed through use? There are clues. For one, if all Neandertals—men, women, and children, especially young children—have a particular feature, all of them would have had to be doing the same level of exertion and activity. And if Neandertals from different environments, for example Europe and the Middle East, all have it as well, then all Neandertals, everywhere, would have had to have the same level of exertion. A simpler solution in these cases is to conclude that the feature in question, say the barrel chest, was determined by genes.

Genes

We actually know some very specific things about Neandertal genes because science has been able to recover many of the genes themselves from bones that are not completely fossilized. Fossilization is a process in which minerals in the soil replace chemicals in the bone (or, rarely, soft tissues), literally turning the bone to stone. But it is a process that can take thousands of years, and for many Neandertal skeletons complete fossilization has not occurred; there is still organic material in the bones. With the proper techniques, paleoanthropologists can extract DNA from this material, actually two kinds of DNA. The most abundant kind, and the easiest to recover, is not the DNA that controls our inheritance, which is found in the nucleus of every cell. It is the DNA found in the mitochondria, which are located outside of the nucleus in a cell's cytoplasm (hence mitochondrial DNA, or mtDNA). Mitochondria are organic structures in the cell (organelles) that help cells produce energy, and each mitochondrion has its own DNA that regulates its function. The curious thing about mitochondria is that sperm do not have them. All of your mitochondria came, ultimately, from your mother, and hers from her mother, and so on. Of course, men do have mitochondria; they just don't pass theirs on to their children. Sexual reproduction does not alter the DNA of mitochondria (unlike nuclear DNA, which is reshuffled during the formation of eggs or sperm), so your mitochondrial DNA is identical to that of your mother's, *unless there has been a mutation.* Mutations are nondirectional, chance changes to the DNA. Sometimes a mutation in mitochondrial DNA renders that mitochondrion unviable so that it no longer processes energy effectively, but more often a mutation has no physiological effect at all and is passed along with the mitochondria to a woman's sons and daughters, and her daughter's sons and daughters, and so on. If two daughters of the same woman go their separate ways and each successfully reproduces, and their daughters successfully reproduce, and so on, over time the two related lines will accumulate different sets of mutations in their mtDNA, and the longer they are apart, the more different their mtDNA will be. When we compare the mtDNA recovered from Neandertal skeletons to that of modern people, we find about three times as many differences (twenty-seven mutations versus eight) as we find when we compare the two most different modern people with one another. This tells us a couple of interesting things. First, Neandertals were at least three times more different from us, in terms of mtDNA, than any of us is from any other modern person. Second, if we apply a rate of change—that is, a rate at which mutations occur—we can estimate how long ago we and Neandertals separated. As you might guess, calculating a mutation rate is tricky, and there is much disagreement. But most paleoanthropologists are comfortable with a rate of change that places the split about 500,000 years ago.[7]

In 2010 Svente Pääbo and fifty-five colleagues of the Neandertal Genome Analysis Consortium published the first draft of their attempt to document the entire Neandertal genome ("genetic" DNA in the cell nucleus). They based their analysis on archaic DNA recovered from the bones of three Neandertal women who died in Croatia about 45,000 years ago, which they compared to the DNA of modern people living in Europe, Asia, and Africa. The results were a bit of a surprise. The primary result demonstrated that 1 to 4 percent of the genes in modern Asians and Europeans had a Neandertal origin. The figure of 1 to 4 percent may not sound like much, but it confirms a suspicion held by many paleoanthropologists that Neandertals and modern humans could, and did, interbreed. Two other results were equally interesting: modern Africans have no Neandertal genes in their genome, and the Asian genome had just as many Neandertal genes as the European genome. The most likely explanation for this state of affairs is that the interbreeding occurred *after* modern humans had migrated out of Africa but *before* they dispersed into Europe and Asia. The most likely time frame for this, according to Pääbo and colleagues, was 60,000 to 80,000 years ago. This range of dates presents a paleoanthropological puzzle. There are Neandertal remains that date to this time frame in the Middle East, but as yet no anatomically modern remains. There are modern remains earlier than 80,000, and then again after 60,000. Given the imprecision of genetic dating by assumed mutation rates, this inconsistency should perhaps not worry us. What is important is that there was gene flow between Neandertals and our ancestors prior to our expansion into Europe and East Asia. But this gene flow, the term geneticists use for the transfer of genes from one population to another, was one-way; no specifically modern human genes were identified in the Neandertal genome.[8] This apparent one-way gene flow inspires questions about how it actually played out in the lives of Neandertals and modern humans. How did they interact? Were there mixed groups? Or is violent interaction more likely? We will return to these issues in the final chapter.

Pääbo's analysis was also able to identify seventy-eight genes found in the modern genome that were not found in the three Neandertal women. It is important to keep this in perspective; the Neandertal and modern genomes were 99.84 percent identical. But we have seen that Neandertals did differ from moderns anatomically, and that some of this difference was probably genetic. Some of the seventy-eight unique modern genes have been linked to known functions, but we must be cautious here. As we have seen, gene expression in anatomy is a complex process that involves the gene itself, other regulatory genes, and an individual's developmental environment. Curiously, of the seventy-eight genes in question, the biggest DNA differences were in genes linked to the skin. And some of the other different genes are those involved in energy metabolism, including a set that leads to Type 2 diabetes if a faulty variety is present. But for our purposes in this book, the most interesting differences are in genes important in cognitive development. Modern humans have at least a few genes linked to cognitive development that Neandertals did not have. Unfortunately it is not entirely clear what these genes do. There were also some curiosities. One of the Neandertals tested by Pääbo had the gene for red hair, for example. This discovery was a surprise, but perhaps it shouldn't have been. The gene for red hair is related to genes controlling skin pigment, and anthropologists have long known that reduced UV radiation in high latitudes (e.g., Europe) selects for reduced skin pigmentation. Neandertals had lived in high latitudes for several hundred thousand years, far longer than modern Europeans, so evidence of reduced pigmentation should surprise no one. A more provocative gene identified for Neandertals is known as FOXP2. A variant of the FOXP2 gene is known to cause a speech impediment and difficulties in learning and using certain kinds of grammar. Neandertals have the normal variety. Does this mean that Neandertals had modern speech, just like you and me? Perhaps, but not necessarily. FOXP2 is not a gene *for* language or speech; rather it is a gene that is involved in modern speech. There are undoubtedly many other genes involved in modern spoken language that have not been identified. Neandertals may or may not have had them. But their possession of FOXP2 is certainly provocative.

At the end of the day, knowing a few, or even many Neandertal genes does not help us much in our quest to understand their lives, at least not yet. Archaic DNA from Neandertals is most useful for documenting how closely related we are to them and for estimating how long we have been apart. There are not single genes for brow ridges, or large noses, or barrel chests—at least not that we know of. Such features result from the interaction and timing of many genes over the course of an individual's lifetime, especially when we are young. So even when paleoanthropologists have described the entire Neandertal genome (and it is only a matter of time), most of our questions about Neandertal life will still remain. How, then, can we understand their unusual appearance?

Out in the Cold

Christian Bergmann and Joel Allen were nineteenth-century biologists (one German, one American) who are now forever linked as the discoverers of two "rules" governing the body shapes of mammals. Bergmann's rule states that for varieties of the same species, body mass will increase as latitude increases: those living in cold climates will have compact, massive, bodies, and those living in hot climates will have more linear body forms. It's simple physics. The ideal shape for retaining heat is a sphere; it has the least surface area to mass.

Mammals are endothermic, which means that they generate their own body heat, so in cold climates individuals who have a more spherical body shape retain heat better, use less energy to maintain body temperature, and reproduce better as a consequence. It's not only simple physics; it's simple natural selection. Allen's rule states that mammals living in cold climates will have shorter limbs relative to torso length than those who live in warm climates. Because arms and legs are narrow and linear, they lose heat faster than the torso, so shorter arms and legs lose less heat than long arms and legs. Modern humans follow both rules. The Inuit of northern Canada have short, compact bodies, with relatively short arms, and Sudanese from tropical Africa are tall and thin with relatively long arms and legs.

Neandertals also followed the rules. They evolved a classic body type for retaining heat: relatively short and compact, with short arms and legs. This explanation is so compelling that paleoanthropologists have tried to incorporate other Neandertal features into it, arguing that Neandertal noses evolved to warm and moisten cold air. However, recent research points to heavy chewing as a more likely explanation for the unusual features of the Neandertal face. The alert reader will also raise an objection concerning cold adaptation for Neandertal bodies. Few modern Europeans resemble Inuit. Indeed if one considers northern Europeans such as the Dutch, a more tropical body type would be a better match. More important, Europe is not a particularly cold place, certainly not arctic. So why should Neandertals have had body shapes like those of modern Inuit? The answer is that Europe hasn't always been as warm as it is today.

Ice Age

Global warming is on everyone's mind these days, at least the minds of those paying attention to world affairs. Most scholars now accept that this warming is caused by human actions. There are still a few skeptics (though very, very few among specialists on climate), who prefer to argue that the recent warming is just the latest in a long series of natural climate changes that the Earth has been experiencing for the past million years or so. At first consideration, they have a point. Earth's climate has been on a roller coaster for over a million and a half years, and if anything, the hills and valleys of this roller coaster—the times of extreme cold and warmth—have gotten more dramatic as the ride approached the present. We are currently in a trough of warm conditions, and have been for about 10,000 years. (And if past timing is any indication, and it is, we are due for cold, not dramatic warming.) But as recently as 20,000 years ago Europe and North America were locked in an Ice Age. Immense continental-scale glaciers thousands of feet thick formed over Scandinavia and Hudson Bay, and the weight of ice pushed their margins as far south as central Germany and central Illinois. These huge volumes of ice had effects far beyond their margins. Intense cold over the glaciers caused high atmospheric pressure ridges to form, which in turn pushed the "normal" climatic zones to the south. As a result much of Europe was not only colder, but it was drier as well. The water that fell as snow over the glaciers stayed there, and ocean levels all over the world fell as much as 120 meters (400 feet), changing the shapes of continents. About 20,000 years ago you could have walked from Paris to London, or Alaska to Siberia. The European hardwood forests disappeared, replaced by treeless cold steppe terrain, which abutted the glaciers, scattering conifers farther to the south. The animals also changed. Instead of woodland mammals such as deer and elk, there were cold grassland mammals such as reindeer, musk ox, woolly mammoth, and woolly rhinoceros. Even the predators were different, with lions and hyenas added to the usual stock of wolves and bears. Europe was a very different place during an Ice Age.[9,10]

By 20,000 years ago Neandertals were long gone, but this most recent major ice age was just the last in a long series. It lasted from about 30,000 to 12,000 years ago. A slightly less intense glaciation preceded it between about 70,000 and 60,000 years ago, and Neandertals were definitely around for that one. (Indeed many of our best preserved European Neandertals date to about this time.) Between 128,000 years ago and 180,000 years ago there was a very long, intense Ice Age, and it was during this one that the "classic" Neandertal anatomy evolved. Prior to 180,000 years ago there were several other Ice Ages, going all the way back to the time of Neandertals' immediate ancestor, *Homo heidelbergensis*. These Ice Ages, and the intervening warm periods, were not uniformly cold or warm; climate during each period fluctuated wildly, from cold extremes to warm extremes, often over intervals of only a couple of thousand years. The average temperature during Ice Ages was lower than the modern average, and within each there were colder intervals and brief episodes of extreme cold lasting only a few thousand years. We know that Neandertals remained in Europe during the Ice Ages, but the archaeological evidence suggests that they abandoned many areas during the episodes of extreme cold; of course, so did modern people when they finally arrived 40,000 years ago. What is clear is that Neandertals evolved in Europe during a time of fluctuating climates that included several long, cold phases that lasted for tens of thousands of years, plenty of time for natural selection to follow Bergmann's and Allen's rules.

So how cold-adapted were Neandertals? The anatomical evidence suggests that they were perhaps not much more cold-adapted than cold-adapted modern people such as the Inuit. Leslie Aiello and Peter Wheeler have done a sophisticated analysis of the Neandertal body type in order to determine just how much cold a Neandertal could tolerate.[11] Using knowledge of Neandertal build and muscularity (derived from skeletal remains) and estimates of their basal metabolic rate that took into account their heavy reliance on meat (more on this later), they were able to determine a "minimum sustainable ambient temperature," which is the lowest air temperature one could survive if naked to the elements, that is, without clothing or fire. The result was 8° C (46.5° F). The same computation for modern Europeans yielded 10.5° C (50.9° F). This is only a small advantage. It tells us that Neandertals were probably more cold-adapted than modern Europeans, but not by much. They would have been unable to live in Ice Age Europe without the aid of cultural mechanisms for keeping warm: fire and clothing.

Neandertal noses have also been implicated in cold adaptation. Modern people who live in cold climates tend to have long, high noses. Anthropologists think that this feature evolved to warm and moisten air a bit more before it entered the lungs. But modern long, high noses are also narrow, not wide like the noses of Neandertals. Some anthropologists think that for Neandertals, nose size may instead be linked to their high level of exertion, during which an ability to take in and breathe out larger volumes of air through the nose might be an advantage.[12]

It does seem, then, that cold adaptation can explain *some* of Neandertals' distinctive appearance, but far from all of it. The shape of the cranium and brain, the protruding midface, the lack of a protruding chin, and the general robusticity of the bones are not cold adaptations. So what accounts for them? How did they help Neandertals adapt? As it turns out, we may be asking the wrong question. In the features just mentioned, Neandertals actually resemble many prehistoric hominins, including *Homo heidelbergensis*, their immediate ancestors on the family tree. These features are in fact normal features for nonmodern hominins and were true of archaic forms in Africa and Asia as well as Europe. It is we modern humans who are different.

Thus far we have contrasted Neandertal anatomy with our own and tried to explain Neandertals' unusual features. But most are not unusual when compared to other hominins of the same age or older. Our perspective, taken from our own anatomy, has blinded us not to the oddity of Neandertals, but to the oddity of ourselves. Erik Trinkaus, an authority on Neandertals we will encounter often in this book, set out in 2006 to compile as definitive a list

of distinctive Neandertal traits as possible. He identified seventy-five possible traits from the immense literature on Neandertal and modern human
anatomy, and then used sophisticated statistical methods to identify those
that did the best job of separating Neandertals from moderns. He found that
only 25 percent could reasonably be considered distinctive for Neandertals.
Twice as many, 50 percent, were distinctive for moderns. The remaining 25
percent were shared. Clearly, it is *Homo sapiens sapiens* who are the more
distinctive anatomically.[5]

Neandertal Brains

This is a book about Neandertal ways of thinking, and the anatomy most
involved in thinking is, of course, the brain. For centuries people have known
that the brain is the organ of thought, but it has only been in the past few
decades that scientists have begun to understand just how the brain operates.
And it is not simple. About 150 years ago scientists began discovering things
about the human brain that would stand the test of time. By about 1860
it was known that the outside and top parts of the brain, the cortex, had a
major groove (the longitudinal fissure) from front to back that divided the
brain into two hemispheres. Scientists suspected that the left hemisphere
was more responsible for major speech functions than the right hemisphere.
It was known that the two hemispheres communicated through the corpus
callosum. (In fact in "split-brain" patients, it is the corpus callosum that is
cut, a procedure that is still done to reduce the severity of epileptic seizures.)
Scientists also came to understand that the right part of the cortex seemed to
control the left side of the body, and vice versa. It also appeared that the front
part of the brain (left and right frontal lobes) was responsible for our reasoning
abilities, imagination, planning, and decision making. (Destroying small
amounts of brain tissue in each frontal lobe is known as a frontal lobotomy; it
is still done—voluntarily!—to reduce extreme aggression in mental patients.)
It was known that the brain stem, where the spinal cord joins the brain, is the
area responsible for our basic vital functions, such as heart rate, breathing,
and blood pressure, and that these functions can continue to work without
the cortex (which is why some patients are said to be "brain-dead," yet they
continue to breathe on their own, even in a coma: in these patients, the cortex
is considered "dead," not the brain stem). It was also known that the cortex
has a minor groove (central sulcus) running from ear to ear, and this strip
of the cortex in front of this groove appeared to be responsible for physical
movements (this strip is called the motor cortex), while the surface behind

the groove appeared to be responsible for integrating our physical senses with the external world, particularly our sense of touch (this strip is called the sensory cortex, and the whole area is called the parietal lobes).

We have no living Neandertal brains; indeed we have no Neandertal brains whatsoever. (Believe it or not, in unusual circumstances brains can fossilize, or at least the external shape can be preserved, and anthropologists even have one for a much earlier hominin.) But this does not mean that we are completely ignorant. We do have Neandertal brain cases, from which we can determine overall brain size and also overall brain shape. As an individual matures from infancy to adulthood, the upper part of the brain case expands as the growing brain exerts pressure from the inside. In a sense, the brain pushes out the cranial vault, and as a result the internal shape of the brain case preserves the shape of the brain. Neandertal brains did differ from modern brains in both size and shape. Let's first look at size.

Neandertal brains were bigger than modern brains, about 10 percent bigger. Modern brains average about 1,300 cubic centimeters; Neandertals averaged 1,427 cubic centimeters. We must be cautious here. We do not have a large number of Neandertal crania, only twenty-eight complete enough for a reliable measurement of volume, and it is possible that by chance we have found some big ones. (The Amud Neandertal had a cranial capacity of 1,740 cubic centimeters, about 400 cubic centimeters larger than the modern average.) Also Neandertals were heavily muscled, and some of the increase in brain size may reflect their body size, though given their short stature and overall weight (180 to 200 pounds) this strikes us as unlikely. Most anthropologists accept that Neandertals had slightly larger brains, on average, than modern people. But were they 10 percent smarter? We don't think so. For one thing, there is no clear correlation between brain size and intelligence (however you want to measure it) in modern people. For another, raw brain volume may not accurately reflect the number of neurons, or the size of any other functional unit of the brain. It is just size.

Of course, when we compare animals to one another there *is* a correlation between relative brain size and performance on intelligence tests. But we have to factor out body size, because large animals need larger brains to monitor and run their large organs and higher number of nerve endings. Any measure of brain size must take into account the size of the body. When we factor out body size, some animals (fish, reptiles) still have very small brains, and others (mammals, birds) have large brains. As you might predict, fish and reptiles do poorly on intelligence tests. (TW's son owned a pet lizard that learned just one thing in its lifetime: When the lid of the aquarium opens, crickets fall from the sky. Mice would learn that in about two trials.) When we compare mammals to one another, some have relatively small brains (horses, moles), and others large brains (primates, dolphins), and even here there appears to be some correlation with intelligence. But when we compare primates to one another, the size = intelligence equation is harder to justify. All are relatively large-brained, and all are pretty smart. (In India, for example, teams of monkeys work together to distract and rob people on the streets.) But there is one primate that *is* very different: humans. Our brain is about three times bigger than it should be for our body size, and we can devise tests that no other primate can pass. So brain size is not irrelevant. Moreover brain size comes with a very high cost. About 20 percent of our metabolism—our caloric intake, if you will—is devoted to keeping our brain running.[13]

What about Neandertals? They had large brains, which on a coarse level might suggest that they were smarter than we. Is there any evidence to suggest that this was true? Most anthropologists conclude, based on archaeological remains, that Neandertals were *neither* more nor less intelligent than we. In this book we will reach a slightly different conclusion: Neandertals were *both* more *and* less intelligent. In other words, the question is not very helpful. Neandertals were different intellectually from us—not more or less, just different. Brain size doesn't tell us this. Our first clue comes from brain shape.

Neandertal brains were longer and broader than ours, and were broadest in the middle when viewed from behind.[14] Ours is broadest high on the sides. Their frontal lobes had about the same volume as ours, which you may find surprising if you know a little about brains. The frontal lobes, which are located in the front of the brain behind the eyes and forehead, control many of the so-called higher brain functions such as reasoning, planning, and emotional control. So if Neandertals had frontal lobes as large as ours, they must have been as adept at reasoning, planning, and emotional control as modern people, right? It is not quite that easy. The equivalence in size certainly suggests that Neandertals were unlikely to have been very different, but it does not require equivalency. The various lobes of the brain are interconnected in complex ways, and differences elsewhere in the brain cannot be ignored. The volume of Neandertals' parietal lobes was smaller than ours.[15] These lobes are high on the sides and back of the brain and are associated with a truly bewildering array of cognitive systems and abilities, from perception to spatial thinking to understanding speech to creating mental models of the world. At our current of state of understanding we just cannot take this bit of data—Neandertals having smaller parietal lobes—and come up with even a single likely difference in behavior. This is why archaeological remains are far more important for interpreting Neandertal thinking and will form the basis for most of our presentation.

The Tough Life

Shanidar #1 lived and died in Iraqi Kurdistan about 50,000 years ago. He is famous, for a Neandertal. Not only is he well-known among paleoanthropologists as one of the most complete Neandertal skeletons, but he has acquired a literary life of his own. He was the model for Creb in Jean Auel's *Clan of the Cave Bear*. This fame is based largely on details of his personal biography, for Shanidar #1 led a tough life. We know this from his bones. Like the glamorized forensic anthropologists featured on popular TV dramas, paleoanthropologists are in fact able to reconstruct details of a fossil individual's life story. Many of these details are mundane: how old the individual was, what sex, how tall, how heavy, and so on. Shanidar #1's pelvis tells us that he was male (it is narrower than a female's); the sutures where the bones of his skull meet tell us that he was between 30 and 40 years old when he died (more likely toward the 40 end); and his arm and leg bones indicate that he would have stood about 5 feet 8 inches, one of the tallest Neandertals known. Nothing particularly exotic here. But paleoanthropologists, in this case Erik Trinkaus, are able to examine bones and identify physical traumas—injuries or disease—that occurred during life.[16]

And Shanidar #1 was a mess. His right arm had been entirely useless for many years prior to his death; indeed it appears that he may actually have lost his entire forearm: none of the bones of the right forearm were found with the skeleton, and the healed condition of the end of the upper arm (distal humerus) suggests amputation just above the elbow, either by direct injury or perhaps by another individual who removed the mangled limb. The upper right arm, the shoulder blade, and the collar bone are about 10 percent smaller than the left, and the surface layers of the bones are much thinner. This could have been the result of interrupted growth due to paralysis as a child, or the posttraumatic atrophy of the bones as an adult. The trauma also affected his right foot and leg. He had broken one of the bones of the arch of his right foot (the one leading to his pinky toe), and the break had led to arthritic degeneration of the bones of the right ankle, probably because Shanidar #1 had to alter the way he walked because of the break. The effect extended to his knee, which shows serious arthritic degeneration. The left foot, leg, and knee are normal. Apparently something gruesome had happened to Shanidar #1's right side.

And if an almost useless right side were not enough, Shanidar #1 had suffered a crushing blow to the left side of his upper face, a blow that fractured the orbit, probably blinding him in that eye, and also the left cheekbone and left side of his cranium. Like the injury on his right side, this facial trauma had entirely healed long before Shanidar #1 died. Finally, he had suffered a separate cutting wound to his right scalp, deep enough to cut the bone. He survived this as well.

Any of these injuries, excepting perhaps the broken foot, could easily have killed Shanidar #1, but they didn't. How did they happen? There are a number of possibilities. Trinkaus, in cautious scholarly style, has suggested three possible scenarios that would account for the evidence:

1. Shanidar #1 suffered an accident, perhaps a rockfall from the ceiling of Shanidar cave, that crushed his right side, and perhaps also his face.
2. The blow to his face and head also injured the left motor cortex of his brain, which controlled the right side of his body. This neural damage produced right-side paralysis and, ultimately, atrophy of his arm. (We find this one unlikely. A blow to the left side of the head is more likely to cause damage to the right side of the brain when it crashes into the right skull wall, something known as contre-coup effect.)
3. Shanidar #1 received a "traction" injury to the right shoulder (his arm was "pulled from its socket"), resulting in a paralyzed shoulder and arm with subsequent atrophy. The right foot and left face injuries were separate events.

What is perhaps most significant is that Shanidar #1 survived, not for a few days, but for years. He could not have done this alone. Someone nursed him to health, and after his recovery he continued to participate in Neandertal life. This is our first clue to the working of the Neandertal mind.

It is likely that Shanidar #1's crushed right side resulted from an accident. His facial injury, however, is less clearly accidental, and indeed could have resulted from a heavy blow to the head delivered by someone else. Paleoanthropologists are reluctant to attribute interpersonal violence to prehistoric people unless there is clear evidence, but it almost certainly happened. After all, violent interaction is characteristic of not just modern humans, but of many modern nonhuman primates, including our nearest cousins, the chimpanzees. It would be an overreaction to deny interpersonal violence to Neandertals just because it fits an unfortunate stereotype. If Shanidar #1 was injured in a fight, he was not alone. Shanidar #3 was murdered.

The skeleton of Shanidar #3, who was not a direct contemporary of #1, is not as well preserved, but his ribcage is in good shape, and it has a forensic tale to tell that is worthy of *CSI*. The upper edge of the left ninth rib (counting down from the top) has a deep gash cutting through to almost the middle of the bone. The only object in Neandertal life capable of producing the wound was a stone tool, either the edge of a knife or, more likely, a spear point. The edges of the gash had begun to heal, but Shanidar #3 died before the healing was complete. The center of the cut bone had *not* begun to heal, however, which suggests that the spear point had remained in the wound until Shanidar #3 died. Trinkaus estimates that Shanidar #3 survived for several weeks before succumbing to his wound, perhaps from massive infection. The spear point had probably penetrated the chest cavity to the pleural lining of the lung, and although such a wound would not necessarily have been fatal, it was in this case. Some anthropologists, repulsed by violence perhaps, have suggested that Shanidar #3 had just been clumsy and had fallen on his spear, a kind of Inspector Clouseau in Neandertal garb. This strikes us as farfetched. It is far more likely that someone stabbed him. But who? Steven Churchill has recently reexamined the famous ninth rib and concluded that the wound was produced by a thrown spear, not the usual Neandertal thrusting spear.[17] Churchill performed an experiment using pig carcasses (they are similar to human rib cages in size and muscularity) and reconstructed Neandertal spears. These spears produced far more damage to the pig ribs than seen on Shanidar #3's rib. But when he used a modern human spear propelled by a spear thrower, the result closely resembled Shanidar #3's wound. As far as we know, Neandertals never used spear throwers. Thus not only was Shanidar #3 killed, but he may have been killed by a modern human. Archaeological remains suggest that modern humans entered the Middle East soon after 50,000 years ago (there are no skeletons of this age), so the scenario is at least possible, but many anthropologists demur, preferring to avoid attribution of Neandertal-modern aggression based on a single enigmatic example. But it does raise the inevitable question: Just how long have people in the Middle East been killing one another over irreconcilable differences?

As melodramatic as this scenario is, it serves primarily to reemphasize our current theme. The Shanidar Neandertals led tough, dangerous lives, in which serious injury was common and expected. Of the six reasonably complete adult Neandertals excavated from Shanidar, four had suffered serious injury and survived. And *all* of the older Neandertals (elderly being 35 to 40) had suffered injury. It was a predictable part of their lives. Similarly all of the adult Neandertals at Shanidar suffered from degenerative joint disease in the knees, in the shoulders and elbows, in the ankles and feet, and in the back. Mostly this was not debilitating, except in cases like Shanidar #1's, which were indirect consequences of injury. But it was painful. The Shanidar Neandertals acquired their arthritic joints from constant, heavy exercise. And all died by the time they were 45. Neandertal life was nothing to aspire to.

The Shanidar Neandertals were not an aberration. The pattern of injury and degenerative joint disease presented by their skeletons is repeated wherever Neandertals are found. The "Old Man of La Chapelle-aux-Saints," perhaps the single most famous Neandertal of all, suffered from severe arthritis, had a degenerative hip socket and a healed broken rib, and had lost most of his teeth, including all of his chewing teeth. And when he died some 50,000 years ago he was not all that old, probably between 30 and 40. La Chapelle's traumas, tooth loss, and degenerative conditions so affected his anatomy that they hindered an accurate reconstruction. In 1910 the French anatomist Marcellin Boule described him as hunched over and unable to straighten his legs, an image that immediately became popular and is largely responsible for the mistaken image of Neandertals that most people carry around in their head.

A more recent image of Neandertals is more accurate, but startling: a rodeo cowboy. Erik Trinkaus has long been intrigued by the pattern of injuries observed on Neandertal skeletons (not surprising, given his detailed study of the Shanidar skeletons), and in the early 1990s he and Thomas Berger set out to see if they could find any similar patterns among modern human skeletons. They examined seven groups of skeletons: three samples of modern people, three archaeological samples of prehistoric people, and one sample of athletic injuries among rodeo cowboys. If you have seen or been to a rodeo, you know that the competitive events pit people against large animals: horses, cattle, and most dangerous of all, bulls that have been bred specifically for their ability to throw off young men who try to ride them. Injuries are common, so common that the Professional Rodeo Cowboys Association kept track of all of the injuries received by its members between 1981 and 1990, finding 2,593 in all. This sample turned out to be the eye-opener for Berger and Trinkaus.[18] The pattern of injuries found on Neandertal skeletons was very similar to the pattern found in the rodeo sample: "There are little more than trivial differences between the various Neandertal lesion tabulations and the rodeo one." In both samples there was a high incidence of head and neck injuries, twice that of the "normal" sample. Berger and Trinkaus concluded that Neandertals must have had "close encounters of a nasty kind" with large beasts. Most of the injuries appear to have occurred during Neandertals' prime years, that is, as young adults. And 87 percent of the Neandertal head and neck injuries occurred on males, a fact to which we must return in a later chapter. We do not think that Neandertals rode bulls or drove beat-up pickups listening to old Willie Nelson recordings, but we do think a regular part of their lives was a brutal kind of hunting dangerous enough that serious physical injury was predictable for those who survived into adulthood.

Not only did Neandertals lead dangerous lives with high levels of physical exertion, but they also faced periodic food shortages. This is another tale told by their skeletons, in this case by their teeth. Permanent teeth begin to form in the jaw soon after birth. Enamel is deposited from a growth center at what will become the crown of the tooth; the process produces a sequence of enamel layers running down the side of the tooth, with the older layers on the bottom. If an infant suffers a period of poor nutrition, the deposition of enamel will be disrupted, and this disruption will be permanently preserved as pits or grooves in the adult tooth, a condition known as hypoplasia. There have been several studies of Neandertal teeth that looked for evidence of such hypoplasia, and all found it in significant frequencies. Debbie Guatelli-Steinberg and her colleagues identified it in 39 percent of the Neandertals they examined, a figure that agrees well with other studies. In other words, almost 40 percent of Neandertals suffered significant periods of poor nutrition as infants.[19] And since they were almost certainly nursing at that point in their lives, it also means that their mothers were under stress; hunger appears to have been common. But we must place this number in context before we can conclude that Neandertals were poor providers. When Guatelli-Steinberg applied the same techniques to prehistoric Inuit jaws from northern Canada that were a few hundred to a few thousand years old, she found the numbers to be the same: 38 percent presented evidence for hypoplasia. So the message here is not that Neandertals were inept hunters and gatherers, but that hunting and gathering in cold, hostile environments is always a risky, unpredictable affair. As we will see in the following chapters, however, Neandertals were not as technologically sophisticated as the Inuit; that they were able to achieve comparable levels of nutrition with simpler tools is a testament to the success of their more physical, dangerous approach to daily life.

Cognition

What does our grim portrait of Neandertals reveal about their thinking? Did their tough, dangerous lives favor certain ways of seeing the world or certain personalities? We will introduce much more information about Neandertal lives in the chapters that follow, and many of their ways of thinking will emerge. At this point we offer only a few generalizations about Neandertal lives that have implications for their thought processes:

1. Neandertals had the ability to withstand the pain, discomfort, fatigue, and hunger that were part of their everyday existence.
2. Death was a common companion; Neandertals faced their mortality every day. Later we will demonstrate that Neandertals had a concept of death, and argue that it was clearly something that they thought about.
3. With a high mortality rate and few survivors past the age of forty, Neandertal families extended to no more than three generations.
4. They knew interpersonal violence.
5. But they also showed compassion by caring for disabled individuals.

These five generalizations are not much to go on, but they are a beginning. From them we can infer a few cognitive and personality traits:

1. Tenacity, or dogged persistence. Neandertals must have been able to complete their tasks while in pain or with diminished capacity.
2. Wariness. They must have been wary by nature, and probably intolerant of strangers. It was the only safe way to interact. We will build on this point in later chapters.
3. Love. They formed emotional bonds with family and band mates. Someone nursed Shanidar #1 back to health, and someone strived for weeks to keep Shanidar #3 alive, to no avail. Given the harsh conditions of Neandertal life, this is evidence of a strong emotional attachment.

These traits were not unique to Neandertals. You and I have them too. They are basic thought processes that we share, and they attest to our common humanity. And so we must look elsewhere for differences.

Further Reading

1. Arsuaga, J. L. 2002. *The Neanderthal's Necklace: In Search of the First Thinkers.* Trans. A. Klat. New York: Four Walls Eight Windows.
2. Stringer, C. B., and C. Gamble. 1993. *In Search of the Neanderthals: Solving the Puzzle of Human Origins.* New York: Thames and Hudson.
3. Wolpoff, M. 1999. *Paleoanthropology.* Boston: McGraw-Hill.

4. Trinkaus, E., and P. Shipman P. 1992. *The Neandertals: Changing the Image of Mankind*. New York: Knopf.

5. Trinkaus, E. 2006. Modern human versus Neandertal evolutionary distinctiveness. *Current Anthropology* 47:597–622.

6. Slatkin, M. 2009. Epigenetic inheritance and the missing heritability problem. *Genetics* 182:845–852.

7. Green, R., et al. 2008. A complete Neandertal mitochondrial genome sequence determined by high-throughput sequencing. *Cell* 134:416–426.

8. Green, R., et al. 2010. A draft sequence of the Neandertal genome. *Science* 328:710–722.

9. Finlayson, C. 2009. *Neanderthals and Modern Humans: An Ecological and Evolutionary Perspective*. New York: Cambridge University Press.

10. Hoffecker, J. 2002. *Desolate Landscapes: Ice-Age Settlement in Eastern Europe*. New Brunswick, N.J.: Rutgers University Press.

11. Aiello, C. L., and P. Wheeler. 2003. Neanderthal thermoregulation and the glacial climate. In *Neanderthals and Modern Humans in the European Landscape during the Last Glaciation: Archaeological Results of the Stage 3 Project*, eds. T. van Andel and W. Davies. Cambridge: McDonald Institute for Archaeological Research.

12. Holton, N., and R. Franciscus. 2008. The paradox of a wide nasal aperture in cold-adapted Neandertals: A causal assessment. *Journal of Human Evolution* 55:942–951.

13. Allman, J. 2000. *Evolving Brains*. New York: Scientific American Library.

14. Holloway, R., D. Broadfield, and M. Yuan. 2004. *The Human Fossil Record*. Vol. 3, *Brain Endocasts: The Paleoneurological Evidence*. New York: Wiley-Liss.

15. Bruner, E. 2004. Geometric morphometrics and paleoneurology: Brain shape evolution in the genus *Homo*. *Journal of Human Evolution* 47:279–303.

16. Trinkaus, E. 1983. *The Shanidar Neandertals*. New York: Academic Press.

17. Churchill, S. 2009. Shanidar 3 Neandertal rib puncture wound and paleolithic weaponry. *Journal of Human Evolution* 57(2): 163–178.

18. Berger, T. D., and E. Trinkaus. 1995. Patterns of trauma among the Neandertals. *Journal of Archaeological Science* 22:841–852.

19. Guatelli-Steinberg, D., D. Reid, and T. Bishop. 2007. Did the enamel of Neandertal anterior teeth grow differently from that of modern humans? *Journal of Human Evolution* 52(1):72–84.

2

The Caveman Diet

The Slaughterhouse

It was carnage, literally. The butchered carcasses of eleven mammoth and three woolly rhinoceros filled the defile, which was little more than an immense crack in a granite headland. The carcasses had been cut apart and the defleshed bones stacked against the wall. Body parts were everywhere; blood, viscera, and bodily fluids made standing difficult. The Neandertal butchers—for they were undoubtedly Neandertals—must have waded through entrails and gore as they systematically processed the remains of this gruesome Ice Age bonanza. Despite the absence of weapons or evidence of actual wounds, circumstances suggest that Neandertal hunters were responsible for the slaughter.

Today La Cotte de St. Brelade is a rocky promontory above a beach on the southern shore of the Isle of Jersey in the English Channel (Figure 2-1). Even during a modern summer it can be a windy, rainy, and cold place. At the beginning of the next to last Ice Age, about 160,000 years ago, the climate was cooler, and a drop in sea level meant that Jersey was not an island, but the end of a peninsula extending west from Normandy. The promontory of La Cotte still marked the abrupt end of a plateau, however, with a cliff plunging 50 meters to the ground below. The eleven mammoth and three rhinoceros apparently took that plunge in one terrifying, chaotic event. There is no other way to explain their presence in a defile at the base of a cliff. It is just not the sort of place either mammoth or rhinoceros would normally frequent. And the condition and proximity of the bones to each other indicate that it was not a gradual accumulation of single falls. Modern elephants do occasionally stampede, and there are historic examples of Africans stampeding elephants and rhinoceros into pit traps (often lined with pointed stakes). Today in Africa rhinoceros do occasionally forage on the fringes of elephant herds, so

FIGURE 2-1 La Cotte de St. Brelade.

finding the two together at La Cotte is not unusual. Something stampeded these animals, terrifying them into a frenzy, so that when the leaders of the stampede reached the cliff edge and tried to stop, their herd mates behind pushed them over. The individual animals were young, but not infants. The fall may have killed some of them outright, but most were probably crippled, thrashing and staggering about at the cliff base, further injuring themselves and anything or anyone foolish enough to approach. This is not how lions and hyenas, both of which were on La Cotte's landscape, usually hunt. But people sometimes do. And the most dangerous carnivore on the landscape at the time was Neandertal.[1]

Diet

An average Neandertal needed somewhere between 3,000 and 5,500 calories per day. Paleoanthropologists calculate this figure using body weight and estimates of levels of physical activity based on the robusticity of bones and the pronounced areas for muscle attachment. And as we saw in Chapter 1, maintaining body heat was a metabolic requirement that elevated the daily calorie requirement. The estimated 3,000 to 5,500 calories is a lot of calories, roughly comparable to that consumed by a professional athlete during training. Neandertals, men and women, needed to average this intake. So how did they do it?

For much of their time in northwestern Eurasia Neandertals lived in a habitat that had little to offer by way of plant foods. The edible plants that are abundant in the tropics and subtropics—fruit, roots, nuts, edible leaves, and so on—are just not as common in northern latitudes. Even among modern hunters and gatherers there is a strong correlation between latitude and diet. Tropical foragers rely heavily on plants, while temperate foragers rely on a mix of plants and animals. Arctic foragers such as the Inuit derive most of their calories from animal products (sea mammals, fish, and terrestrial mammals). Given Neandertals' geographic distribution in northwestern Eurasia we would predict a diet in which animal products were always important. During Ice Ages this would have been especially true. Europe would have been cooler, with temperate forests pushed to the south. In fact during these glacial periods the European habitat would have been very different, to the degree that there are really no modern equivalents. It would have been colder, but unlike in the true Arctic, there would not have been months of darkness during the winter. The daily hours of sunlight would have been the same as today. The result was a landscape that supported more vegetation than is found in the Arctic and subarctic today, especially in the form of grasses and sedges.[2-7] Of course, Neandertals could not eat most of these plants. They could and did eat a few, such as the roots of certain plants and some of the grass seeds, especially if they were cooked. But such resources would have been relatively few for the Neandertals living in northern Europe. Thus just on ecological grounds we would expect animal products to loom large in Neandertals' diet. What has surprised paleoanthropologists was just how large.[8,9]

Archaeologists had unwittingly misled paleoanthropologists about Neandertal diets. Neandertal living sites, often in the mouths of caves or under rock overhangs, presented archaeologists with a mixture of rubbish, mostly stone tools and animal bones. The bones were usually broken up pretty badly, but archaeologists could often identify the species. The bones of medium to large mammals dominated: horses, bison, wild cattle, red deer, and reindeer. Archaeologists made the reasonable inference that these were components of Neandertal diets. Just how Neandertals acquired them, by scavenging or hunting, was a contentious issue for a couple of decades in the late twentieth century, but by about 1990 archaeologists had compiled a picture of opportunistic hunting that most scholars accepted. Then the bone chemistry experts entered the discussion and surprised everyone.

Many Neandertal remains and associated animal bones retain a good deal of organic material. With appropriate laboratory techniques it is possible to identify the chemical makeup of the bone and, through this, identify components of the tested individual's diet. In particular the ratios of certain isotopes of carbon and nitrogen are good indicators of the source of dietary protein. The isotopic signature of a herbivore will appear different from that of a carnivore, and carnivores' isotopic signatures will differ based on the kinds of animals that each consumes (e.g., marine carnivores such as sea lions have a different signature from terrestrial scavengers and terrestrial hunters). When tested, Neandertal bones clearly fell in the carnivore range. No surprise here. And when compared to hyenas, the local scavengers and opportunistic hunters, Neandertals fell much higher "up" on the carnivore scale. They do not appear to have done much scavenging, which was a bit unexpected. But sometimes, if comparative data on local herbivores are good, it is possible to identify which herbivores supplied the majority of animal protein in Neandertal diets. For Neandertals living in northwestern Europe, the two primary sources of animal protein were mammoth and woolly rhinoceros, precisely the animals slaughtered in the cleft at La Cotte.[10,11]

How had archaeologists missed this? Blame the "schlep" effect. Archaeologists had identified the animal body parts that Neandertals had carried back to their living sites. Mammoth and rhinoceros bones were mostly too big to carry. Neandertals almost certainly did carry mammoth meat back to their living sites, but in the form of filleted meat, which left no trace. Archaeologists had succeeded in identifying a significant but secondary source of Neandertal calories.

Together the archaeological and chemical evidence indicates that Neandertals were not just carnivores; they were "top predators," successfully hunting the largest and most dangerous herbivores on the European landscape. We need to understand how they did it, for this is an important clue to their thinking.

Spearing an Elephant

The Mbuti pygmies of the Congo's Ituri Forest hunted elephants with short spears. As you might imagine, it is difficult to kill or even cripple a mature elephant with a single stab of a spear. In one documented instance the Mbuti hunter threw his spear and wounded the elephant in the side. The hunting party then followed its blood trail, wounded it again, and then waited until blood loss and fatigue weakened it to the point that it could be safely killed. The payoff for patience was more meat than the hunters could effectively use. The Mbuti would typically set up a new residential camp, with women and children, near such a carcass to maximize its potential as a food source. But the Mbuti are not an ideal model for Neandertal mammoth hunting. For one thing, in this example the Mbuti used metal-tipped spears, which are thinner, with more penetrating potential than a stone-tipped spear. But more important, the Mbuti hunted in a closed forest habitat where stealth was easier, allowing closer approach and many more opportunities to find an individual elephant isolated from its herd. Isolation is important. Elephants will charge predators if the herd is threatened.

Mammoth were Ice Age elephants, and though there were some anatomical differences (mammoth had much smaller ear flaps, for example) between them and modern African and Asian elephants, there is no reason to think that their social behavior was significantly different. Modern elephants do not typically live in mixed groups that include both mature males and females. Instead females forage and travel with other adult females and their immature offspring. These groups are headed by a mature matriarch, who will aggressively drive away threats and unwanted males. Males live in bachelor groups or as isolated individuals. When females are in heat they tolerate the presence of adult males, but otherwise mixed groups are rare. This social arrangement presents serious obstacles for hunters lacking high-powered rifles. For obvious size reasons, the optimal target would be a juvenile. However, separating a juvenile from its mother, other adult females, and the matriarch would be almost impossible in normal circumstances. Choosing an isolated male had other risks; adult male mammoth were very big and very dangerous. In fact without some way to handicap the mammoth, hunting it may simply have been too risky.

Archaeologists have, however, been able to document one group of successful Stone Age mammoth hunters, the Paleo-Indian hunters who lived on the Plains of North America 11,000 years ago. These modern humans had a more sophisticated technology than Neandertals, but they faced the same challenges. Their spears were composites that included a long fletched shaft mounted with a detachable foreshaft that was tipped with a thin stone projectile point. Paleo-Indians launched these spears with spear throwers, which were hooked shafts that acted as an additional segment to their throwing arms. Using a spear thrower a hunter can dramatically increase the velocity, range, and penetrating power of the spear. But even with this sophisticated equipment, Paleo-Indians apparently avoided hunting mammoth mano a mano in the open field. All of the Paleo-Indian mammoth kill sites relied on some terrain handicap. In one, several mammoth had been immobilized or slowed by the soft ground in a stream bed. In another, several mammoth were apparently trapped in a narrow arroyo. Like all elephants, mammoth had relatively poor eyesight, and, with risk of charging thus reduced or eliminated, the hunters killed the animals using their spear-thrower-launched spears. Paleo-Indians had another advantage unavailable to Neandertals. The mammoth in North

America had not evolved in the presence of human predation, and may not have evolved appropriate avoidance mechanisms.[12]

Paleo-Indians were not Neandertals, but some aspects of their approach to mammoth hunting look consistent with sites such as La Cotte. Like Paleo-Indians, Neandertals apparently relied on landscape features to provide a local advantage. In the case of La Cotte, they not only immobilized the mammoth, but they also injured them with a fall. But unlike Paleo-Indians, Neandertals did not have long-range projectiles and could not stand back at a safe distance launching missiles at the injured mammoth. Instead they used stone-tipped thrusting spears. Is there any wonder that Neandertal hunters often suffered upper body injuries? The use of landscape features suggests that Neandertals were effective tactical hunters, able to assess and deploy any natural advantage available to them. Their reliance on brutal, in-close killing suggests a disregard for personal safety that was beyond the limits of what modern humans typically do. Either this, or they were supremely confident.

Reindeer, Tur, and Gazelle

Neandertals did not hunt mammoth and rhinoceros exclusively. In fact the mammoth bias may have been true only of Neandertal groups living on the Ice Age plains of northwest Europe. Other sites suggest a more eclectic approach to being a top predator.

The Neandertals who hunted at Salzgitter-Lebenstedt in modern Germany were going after reindeer, and not just any reindeer; they focused on prime adult males. Here again we see the use of terrain to handicap prey. In this case, Neandertals maneuvered a reindeer herd, or part of a herd, into a steep, narrow valley to cut off the possibility of escape and then killed them with spears. (Archaeologists have even recovered a unique spear point carved from mammoth bone.) The site has the remains of at least eighty-six individual reindeer. However, they may not all have been killed at once; indeed it is more likely that Neandertals reused the location on several occasions, always in the early autumn (determined by the age of death of the few juveniles and the condition of male antlers). Even though the hunters killed some females and juveniles, they were most interested in the adult males. The juveniles and females were not extensively butchered, but the prime males were. And the Neandertal hunters used the same procedure to butcher every one. For example, they had a technique in which they removed an end of the metatarsal (one of the long lower rear leg bones) and popped off the front surface of the bone, revealing an intact marrow cavity for the entire length of the bone. Marrow is high in fat and very nutritious. Reindeer of course are not as large and dangerous as mammoth and rhinoceros, but the remains at Salzgitter-Lebenstedt suggest a similar approach to hunting: using landscape to gain advantage. Salzgitter-Lebenstedt is at least 100,000 years *more recent* than La Cotte, which suggests that the tactics used by Neandertals in Ice Age Europe remained viable for a long time.[13]

The Caucasus Mountains are at the opposite end of Europe from the sites of La Cotte and Salzgitter-Lebenstedt, yet Neandertals there used a familiar pattern. The site of Ortvale Klde is a rockshelter in the southern foothills of the Caucasus overlooking a river valley, at an elevation of about 530 meters (1,750 feet). Neandertals here focused their energies on the local species of mountain goat, the Caucasian tur. Adult male tur can weigh 200 pounds, the females a bit more than half that. Neandertals hunted both. Here again we see that Neandertals biased their hunting toward prime adults, but this time both males and females. The Caucasian tur has a seasonal migration cycle in which they move to high altitude in the summer and lower altitudes during the winter. Males and females congregate in separate herds, with individual males associating with female herds in the autumn, if they can successfully fight off male rivals. But the tur has a habit that Neandertals capitalized on: they use the same paths year after year. When Ortvale Klde was occupied, about 50,000 years ago, it was located near the lower, winter range of the tur. Neandertals simply needed to find a hiding place near a regular tur migration path and ambush them when they passed.[14]

Thus far we have presented a picture of a Neandertal diet that was exclusively based on meat acquired by hunting, and this is almost certainly misleading. Neandertals who lived in cold, glacial environments of Europe probably did rely heavily, perhaps almost exclusively on meat, as there were few edible plants available in this habitat. But not all Neandertals lived in such extreme conditions. The Ortvale Klde Neandertals almost certainly ate an array of edible plants available in the Southern Caucasus. We don't see this part of the diet because plant remains rarely preserve well in archaeological sites of this antiquity. But there are exceptions. One of these is Kebara Cave in the modern country of Israel.

Kebara is a cave on Mt. Carmel, close to the Mediterranean coast. Neandertals occupied the cave from about 60,000 years ago to about 48,000 years ago, roughly contemporary with Ortvale Klde. The environment was quite different from the Caucasus Mountains or northern Europe. The Kebara Neandertals, like all Neandertals, were adept hunters. Here their focus was primarily on gazelle, which are relatively small for ungulates, indeed much like the tur, and fallow deer. Together gazelle and fallow deer made up about 80 percent of the faunal remains, suggesting a bias in the hunting strategy. Occasionally the Kebara Neandertals also killed wild cattle and boar, which are both potentially dangerous, confirming the Neandertals' willingness to face ferocious prey. They also collected tortoises, which probably were not as ferocious. Even though the climate at Kebara was a bit cooler and drier than it is today, it was still a Mediterranean climate, with a greater variety of potentially edible plant foods. And the Kebara Neandertals did eat them. Here archaeologists were lucky. Many plant remains had been charred by fires Neandertals built in the cave, and charred remains have a much better chance of preservation. Archaeologists recovered more than 4,000 charred seeds, and 75 percent of these were legumes (pod plants, mostly peas). There were also pistachios in significant numbers, and a few acorns. The legumes were (and are) a good source of protein, as are nuts, which also supply fats. It is almost impossible to determine how important these were to the diet of local Neandertals, but it does tell us that when available Neandertals did incorporate plants into their diets. For much of the time Neandertals used Kebara as a winter–spring base camp, a living site, from which they launched foraging trips, returning with their successes, a situation dramatically different from La Cotte but just as typical of the Neandertal way of life.[15]

The Lay of the Land

To be successful, hunters and gatherers must move, and the decisions they make about when, how often, and where to move are important components of their way of life. What they hunt and gather is, of course, an important consideration. In the recent past, for example, some hunters and gatherers developed techniques for exploiting the annual spawning runs of anadromous fish such as salmon. In a few weeks such fishing groups could acquire enough dried and stored fish to last an entire year. They didn't need to move, and lived in permanent, year-round settlements. Other recent hunters and gatherers in tropical Africa relied on a wide variety of small animals and plants and needed to move more often to avoid depleting local resources. But hunter-gatherer mobility is also influenced by social factors, sometimes as simple as interpersonal conflict, and other times to maximize intergroup mating opportunities. We will take up the social considerations in a later chapter. For the present, we need to address how Neandertal hunters moved within their landscapes.

Neandertal archaeological sites consist of refuse left behind by individuals and groups that lingered in a place for a few hours, days, or weeks. Can we somehow link these places together to establish a pattern? The answer is yes, to a certain extent, at least for regions such as western Europe that have been studied intensively. The key is raw material, the kinds of stone selected by Neandertals to use in the making of stone tools (covered in much more detail in Chapter 3). Different kinds of stone—flint, for example—have different structural, textural, color, and even chemical properties, which means that it is often possible to identify the geographic source of the raw material used for a tool or set of tools. This in turn provides a measure of where the makers were going and even a hint at how often.

Neandertals rarely traveled very far for their raw material. In most instances, 80 to 99 percent of the stone could be found within 6 kilometers (3.75 miles) of the site. This is a radius of about an hour's walk. Of course, stone is heavy, and Neandertals almost certainly did not lug it any farther than necessary. This tells us that they knew where to find the best local stone. They could easily acquire it by making a special trip to get it or, more likely, pick some up over the course of other journeys for other purposes. This local scene doesn't tell us much that is surprising; we would expect any large mammal, especially in Ice Age Europe, to be intimately familiar with a home territory of 45 square miles. In many Neandertal sites there is a small percentage of raw material from farther away than 20 kilometers (12.5 miles); occasionally there are small pieces from distant sources, occasionally as far as 250 kilometers, but these are rare. Invariably such stone is of high quality (meaning it is very good for making tools), and it is also heavily reduced, which means that tools made of it were reworked several times until they were too small to be used. This tells us that Neandertal groups did move beyond the local radius of sites regularly but probably not very often. A journey of 20 kilometers is a day's journey—there and back in a day, if one is pressed—and it is less likely that a Neandertal would make a special trip just to get some nice stone. But if he or she passed such a source while journeying for another reason, then carrying some back might be a good choice.[16]

The pattern of raw material acquisition reveals other insights into Neandertal mobility. They almost never traveled across major obstacles such as mountains or large rivers. Most of their local mobility occurred up and down river and stream valleys. Longer journeys, the ones over 20 kilometers, often crossed into adjacent river valleys, usually restricted to environmental zones (and game distributions) that were extensions of the zone occupied in the home valley.

The pattern of Neandertal movement in eastern Europe was a bit different. Most of the raw material was still local, confirming Neandertals' preference for local resources. But when they did journey farther afield the distances were longer, with some raw material traveling over 100 kilometers. Here again it is unlikely that they made special trips to retrieve raw material. Most likely

Neandertal hunters in these colder regions traveled farther in pursuit of game (which may itself have had bigger territories) and occasionally visited sources of high-quality stone. Also possible, though less likely, is that Neandertal residential groups were separated farther in space, and when the occasional spouse exchange occurred, the exogamous individual had farther to travel (i.e., someone, either young men or young women, must have "married out"; more on this later).[17]

Neandertal sites themselves come in three basic types: carcass butchery, small camps, and larger living sites. Some of these sites, usually found in rock-shelters where there is an overhang for protection (which also helped preserve them), are so small that they could accommodate only a few individuals, almost certainly a subgroup of the regular residential group, perhaps a foraging party of some sort camping away from home for a night. The larger residential sites appear to have been central locations that acted as a base of operations. Groups would move out from these on foraging trips and return carrying food that would be shared among the group. Such sites often have the remains of *parts* of prey animals such as horse, bison, or reindeer (tur in the Caucasus, gazelle in the Levant). The smaller carcasses (e.g., gazelle) could have been carried whole, the larger carcasses in movable packages such as hind limbs. But few remains of mammoth or rhinos made it to such places. Instead when Neandertals killed such large beasts the group traveled to the carcass. At a bonanza such as La Cotte, a Neandertal group could feed itself for perhaps weeks, at least until the carcasses became too spoiled to eat. We have no evidence for storage.

Back to La Cotte

We can now revisit La Cotte in light of what we know about Neandertals in general. The Neandertals who butchered the mammoth and rhinoceros at La Cotte were adept hunters, clearly. But we needn't invoke superhuman abilities or strength, or even any specialized skills linked specifically to mammoth hunting. Neandertals had a flexible approach to the food quest. They were able skills adjust to whatever was locally available and could even expand or contract territorial size according to the nature of local resources. The headland at La Cotte was an obvious landmark for the local group that foraged up and down the adjacent river valley (now covered by water). The plateau behind the headland was a place used occasionally by a local mammoth herd. One day a group of Neandertals got very lucky. The mammoth herd, grazing with a few rhinoceros on the periphery, had moved perilously close to the cliff, and they were upwind from the Neandertal hunting party. Somehow the Neandertals startled the herd into a panic; perhaps they had used terrain to sneak up unnoticed, or emerged from a cold fog. However they did it, the ensuing chaos drove eleven mammoth and three rhinoceros over the edge. Then the situation became really dangerous for the Neandertals. They worked their way down to the cliff base to kill the beasts that were still alive. And they killed them with thrusting spears. Most likely they waited until the thrashing had tired the crippled animals. But it was still a dangerous task. Meanwhile others in the party returned to the home site to gather the other members of the community, who then all moved to a spot near the carcasses. The butchery could then begin. If the weather was cold, the butchered carcasses, cut into smaller freezable pieces, might keep quite a while, supporting the entire community. But there was also the problem of scavengers, especially hyenas. There is no evidence that hyenas ever ravaged the bones. Perhaps there were no hyena nearby, though the stench probably carried for miles, attracting all sorts of unwanted guests. More likely Neandertals used fire and vigilance to keep them at bay; there is charcoal in the deposits. And hyenas had undoubtedly evolved to recognize Neandertals as particularly dangerous foes. After days, or even weeks, the Neandertals moved on, perhaps returning to the home base they had previously used, leaving La Cotte to the wind and cold.

What Is Missing from This Picture?

Neandertals were successful hunters and gatherers, but they were unlike any hunter-gatherers living in the modern world, or indeed any living in the past 20,000 years or so. Compared to modern foraging groups, Neandertals relied on a much narrower range of resources, primarily the products of medium-size to large herbivores. Yes, Neandertals did also hunt and eat smaller animals and gather edible plants, but not to the degree true of more recent hunter-gatherers. In environments like the Middle East the archaeological record of more recent, modern hunter-gatherers (within the past 20,000 years) includes abundant evidence for reliance on plants (especially nuts but also some cereals), small mammals (even gerbils), and birds. The small mammal and bird bones preserve well and would have preserved in Neandertal sites such as Kebara had Neandertals used them. Instead we see very little evidence for reliance on small mammals or birds, and the evidence for plants is minimal. The Kebara Neandertals focused on gazelle and fallow deer and scooped up the occasional tortoise on the way home. Higher latitude Neandertals were even more focused on mammals.

Some might argue that Neandertals were like the modern Inuit in this dietary bias. After all, the Inuit rely very heavily on animal products, especially reindeer and sea mammals. But there are significant differences in the two sets of hunter-gatherers. Inuit diets are a bit broader, incorporating sea mammals and occasionally fish, but the major difference is technological. Inuit rely on a sophisticated technology that includes complex gear (e.g., harpoons), boats, sleds, and sewn clothing, to name just a few. Neandertals were pedestrian hunters armed with simple thrusting spears. The Neandertal way of life simply has no analogues in the modern world, and its relative simplicity (in range of foods) has an interesting corollary.

Neandertals probably lacked a significant division of labor based on sex and age. This argument has been developed by the archaeologists Mary Stiner and Steven Kuhn, and though it may strike readers as unlikely at first, it does explain some of the monotony of the Neandertal archaeological record.[18] Among modern hunter-gatherers in tropical and temperate latitudes, women supply much of a group's caloric requirements by gathering plants and small creatures. The men hunt, and although meat from large animals is prized by all, it actually supplies fewer calories over the long term than the foods gathered by the women. Of course, women occasionally hunt and men occasionally gather, but these are not their primary economic roles for hunter-gatherers. And in many hunter-gatherer groups, children assist their mothers in gathering (though their caloric contribution is small). In Arctic groups such as the Inuit, women make and maintain much of the elaborate technology without which the entire group would perish. Archaeologists can easily find evidence of this hunter-gatherer division of labor in the form of plant-processing equipment (e.g., grinding stones), bones of small mammals and birds, or, in the case of Arctic hunter-gatherers, elaborate technologies.

Neandertal sites give us none of this. Either Neandertal women and children did nothing to aid the food quest, which is unlikely, or they were part of the regular hunting strategy. This also makes sense of features of Neandertal anatomy. Neandertals show evidence of heavy exertion even at a young age. If all Neandertals participated in hunting, we would expect this. Neandertal adult males often suffered upper body injuries, but adult females not as much. This suggests that women and children performed different, and less dangerous, tasks during a hunt. Stiner and Kuhn suggest that the women and children could have helped drive animals or beat the bushes to expose the location of game. But even if women and children just helped butcher and transport meat back home, their bodies would tell a tale of heavy exertion. A sexual division of labor is ubiquitous among modern hunter-gatherers;

indeed it is ubiquitous among modern humans in general. This is not a sexist over-generalization, or an assertion of how things should be; it is just a feature of modern human adaptations. Neandertals appear to have been different.

Cognitive Implications

With the evidence for Neandertal hunting and gathering in hand, we can begin to ask questions about Neandertal thinking. How did they organize their economic lives? How did they plan? How did they find their way? What and how did they think about animals? Thoughts do not preserve, of course, at least in the absence of writing or some other recording procedure. Thus everything we write here about Neandertal cognition is the result of inference. The evidence we have reviewed is real. The archeological patterns and anatomical features we have described are not products of our imagination; they are actual clues that have survived from the deep past. Our task is to turn these clues into a picture of Neandertal mental life using what we know from modern psychology and anthropology. In this we will follow a few general guidelines.

First, because Neandertals are genetically and temporally the closest to us of all nonmodern humans, we expect that the similarities between us and them will far outweigh the differences. The default interpretation, the null hypothesis if you will, is that Neandertals and modern humans cannot be distinguished on a cognitive level. Second, we will try not to stray too far from the evidence, archaeological or anatomical. Though it is enlightening to try to fill out the sparse evidence for Neandertals with detailed scenarios (e.g., the fog example at La Cotte), we really cannot then use the scenario as hard evidence for Neandertal cognition. Third, we will base our accounts on well-established understandings of cognition drawn from psychology and anthropology. We will try to avoid commonsense terms such as "intelligence" and "sophistication" in favor of better defined terms. Fourth, we will not unleash our imaginations, a temptation that is quite real when it comes to Neandertals. However, we will not lock away our imaginations entirely. It would also be much too easy to hide behind a dull recitation of facts in order to avoid the inevitable criticism that attempts to present a cognitive profile of anyone. We will occasionally rely on thought experiments, a time-honored technique in which a scholar constructs imaginary scenarios to help explain the interrelationship of complex factors. Our La Cotte scenario is one such thought experiment, based on archaeological evidence fleshed out by what is known about elephants and hunting and gathering in general. When we compare Neandertals to modern humans, we will use a fairly generous range of modern examples, from hunters and gatherers living today to those living in Ice Age Europe 25,000 years ago. We will avoid using archaeological examples whose authorship or status is uncertain or controversial. Our Neandertal examples will also come from various times and places. Our goal is to describe Neandertal cognition to the degree we can and contrast it with ours. We will not try to resolve long-standing controversies in paleoanthropology.

Hunting was not unique to Neandertals, so the mere fact of it tells us almost nothing about Neandertal cognition. There are many ways to hunt. Just think of the differences between wolves and hunting dogs, house cats and lions, cheetahs and hyenas. We have seen that the archaeological evidence points to Neandertals being top predators, so perhaps we should compare them to lions to gain insight into their thinking. But this would not be a good choice. We (and Neandertals) are only distantly related to lions, who are separated from us by at least 65 million years of separate evolution during which the felines selected for a very "hunting-ready" brain and a mind quite different from ours. Neandertals were primates, and some primates do hunt. It is here that we need to begin our inquiry into Neandertal thinking while hunting.

Chimpanzees hunt monkeys, and occasionally other small mammals, and the way they do it presents us with several features also found in modern human hunting and gathering. First, chimpanzees hunt cooperatively; they use the combined action of a group of cooperating individuals to surround and kill prey. Modern human hunters also do this. So did Neandertals. Second, chimpanzee males do most of the hunting. They are bigger and more powerful than females, but this does not appear to be the sole reason; there is also something social going on, having to do with male dominance.[19] This is also true of modern hunter-gatherers, and was probably so for Neandertals. For cognitive purposes, the most important insight provided by chimpanzee hunting is the role of cooperation. How did Neandertals cooperate?

Our second comparison group is modern hunter-gatherers.[20] From what we know about these modern people, and what archaeology tells about Neandertals, we can draw up a list of similarities:

- Cooperative, group hunting
- Use of terrain features
- Carrying meat back to home bases
- Dividing carcasses further (for sharing)
- Cooking

- Varying tactics for different game
- Hunting journeys of more than a day

A comparison to modern hunter-gatherers also enables us to identify differences, some of which may have cognitive implications:

- Neandertals used a narrower range of resources.
- Neandertals used a more dangerous killing technique.
- Neandertals varied prey according to season, but did not schedule within seasons.
- Neandertals used an "out and back" pattern almost exclusively.
- Neandertals traveled shorter distances.
- Neandertals lacked a clear sex difference in foraging patterns.

Finding Their Way

Neandertals needed to be able to locate prey and find their way home again. For most animals, this wayfinding ability is an important piece to the cognitive repertoire; few animals wander aimlessly. Animals have evolved a variety of impressive techniques for navigation; some rely on visual landmarks, some follow scents, and some even use the Earth's magnetic field. But Neandertals were good primates, and primates rely on sight and memory and an understanding of spatial relations, so it is on these that we need to build our account of Neandertal wayfinding.

Think for a moment about how you would give directions to a stranger who intends to visit your home or office. What information would you provide? There are actually two ways you could do it, and the one you choose might depend on what you know about the stranger's knowledge. We will start with the more common of the two. You would first, of course, find out his or her location. Then you would describe a path or route; for example, "Turn left out of the parking lot. Drive to the second traffic signal. Turn right. Drive about four miles. Turn left immediately after the tall bank building. Turn right into the second parking lot. My office is on the first floor of the modernist-style building." We can hear some of you objecting, especially some of the men, saying, "No, I would use cardinal directions, such as 'Turn north' after the bank building." But would you use cardinal directions to a stranger, especially at night? Using cardinal directions is a slightly different approach in that it relies on an external spatial framework, but it is still a system based on describing a sequential path from point to point with decisions made at each point.

The second way to give directions would be for you to describe your location in terms of a permanent spatial framework that the stranger knows and can exploit. "I am at 38° 49.22′ N and 104° 43.91′ W." If the stranger has a GPS device, or even an appropriate map, he or she could find your office. Both ways of giving directions work just fine, but we would guess that most of you would use the first. This approach to wayfinding is known, not surprisingly, as "route following."

Route following, or route knowledge, relies on several abilities: constructing a sequence of points, remembering and describing prominent landmarks, using qualitative decisions about space (turn left; go up; etc.). *None* of this requires a comprehensive image of space that is held in attention—that is, a mental map. Instead, according to the anthropologist Tim Ingold, the traveler or the one giving directions imagines a sequence of views or vistas along the route. These vistas consist of salient objects or scenes that are held in memory and then activated in proper order for the chosen journey. From a narrower cognitive perspective (which Ingold might discourage us from using), the traveler relies on remembered scenes and the ability to string them together and to understand basic qualitative spatial relations, which geometricians sometimes term "topological" (e.g., in/out, left/right/, up/down, etc.). Such route knowledge is in fact a very effective way of moving from point A to point B to point C, and is far and away the most common human solution to wayfinding. But what if you want to go directly back from C to A without wasting the time and effort to backtrack through B? Or what if there are no landmarks?

In April 1916 Ernest Shackleton and five companions sailed a modified lifeboat from Elephant Island near the Antarctic Circle to South Georgia Island, a journey of 1,500 kilometers (800 nautical miles) across the stormiest seas on the planet. There were no landmarks, and except for a few fleeting moments, they could see no stars, or even the sun, all the while battling gale-force winds and strong currents. Of course, they did have a sextant, a compass, and a clock, and they were able to take an occasional sun reading, enabling them to locate their current position reasonably well. Nevertheless it was a remarkable feat of wayfinding. On a smaller scale, but without the sextant and compass, Micronesian sailors from the Caroline Islands were able to sail their outrigger canoes across more than 200 kilometers of open Pacific to make landfalls on small atolls only a few kilometers across and a few meters high. And, reputedly, they could make shorter voyages, at night, after an evening of heavy drinking. How? This kind of wayfinding differs from route following and is known in the academic literature as "survey knowledge."

Such survey knowledge is based on some variety of artificial construct that organizes geographic space into a framework that encompasses all available places (and often unavailable places).

Shackleton and his navigator and crew had all learned the standard Western version of geography, in which every point on the globe has a permanent, two-dimensional relation to every other point. If you know you are at point A and want to travel to point B, it is a relatively simple matter to plot a line connecting the points, and voilà, you have a route. The Micronesian sailors used a sidereal system, in which stars, and the passage of stars across the heavens, provide the external framework of directions for locating points in relation to one another (e.g., Truk lies under the rising point of the star Altair if you depart from Pulawat atoll). Both of these systems are general frameworks that exist conceptually apart from specific, individual routes. The Micronesian navigator need not even see the rising of Altair to use the framework. It existed in his head. (And yes, Micronesian navigators were men.)[21–23]

Scholars of wayfinding once considered route following to be simpler than survey knowledge, and therefore probably more ancient. Should we conclude, then, that Neandertals must have used only route knowledge? It is tempting, and might confirm our preconceptions that Neandertals were primitive. But it is not so simple. As it turns out, both systems of wayfinding develop together in children. One is not the precursor of the other; they are just different. And modern people tend to use both, slipping back and forth between one and the other. Our Micronesian sailors are a good example. When sailing to distant (over-the-horizon) islands, the navigators do use landmarks, mostly underwater reefs and banks that affect the surface conditions of the ocean. At night the navigators literally feel these changing conditions through the motion of the canoe. Indeed the Micronesians try to avoid long journeys on which they must rely solely on their sidereal framework because if they lose track of their present location (because of clouds, storms, etc.) they would become truly lost. Such "mixed" systems are not limited to ocean travel. When traveling cross-country, Native Australians use a system of cardinal directions and landmarks. But these Australians also add a component that for them is the crucial piece. The landscape is covered by locations of symbolic significance tied to the "Dreamtime," which consists of locales and tracks followed by powerful ancestors and creatures. These are the landmarks people use to orient their wayfinding. Interestingly, prior to becoming Christian, the Micronesians also invested much of their knowledge with symbolic, ritual significance, but, being practical sorts, they abandoned most of the complex rituals of sailing once they had no religious motivation.

So what specific cognitive abilities are required for such mixed systems of route knowledge and survey knowledge? Earlier we noted that route knowledge requires memory of vistas, an ability to string such memories into an appropriate sequence, and some basic topological spatial relations. For survey knowledge, another ability must be added, something known as "allocentric perception." Route knowledge can exist entirely from the single viewpoint of the person imagining or explaining the route; it is egocentric. But to imagine and use an independent spatial framework, it is necessary to divorce oneself from one's own viewpoint and see landscapes from a different point of view. And not just another person's; it must be an independent, stable perspective that everyone can use. An example is a bird's-eye view of the landscape. There is some disagreement in cognitive science about just how people do this. Some people do, apparently, create mental maps; that is, they construct some kind of internal visual image of how the landscape appears allocentrically. But others don't. There appears to be a great deal of individual difference. As long as you can imagine and maintain a stable framework, even if it doesn't consist of images, you should be able to create and use survey knowledge of a landscape.

We are finally in a position to ask how Neandertals negotiated their landscapes. We know that they rarely traveled far; most of their journeys took place within a few kilometers of their home bases. We also know that most often they used an out-and-back pattern on their journeys, as opposed to circular patterns. All that we know of Neandertal wayfinding could be easily handled with route knowledge alone. But we also know that Neandertals were capable of allocentric perception. The evidence comes not from hunting and gathering per se, but from tools, which are the central topic of the next chapter. Many Neandertals made stone tools that required an ability to imagine alternative viewpoints. One example is a tool known as a handaxe, which had three-dimensional symmetry. Even the cross-sections of these tools were symmetrical. To conceive and make such cross-sections it is necessary to imagine them; you just can't see into the stone yourself. So we know that Neandertals had allocentric abilities. There is nothing that we know of that would keep them from using it in wayfinding.

Finally, we need to mention sex. Many readers will be aware that there is a significant, and reliable, difference in the scores of men and women on certain tests of spatial cognition. The reason for this difference, and its relevance or lack of relevance to anything else, remain controversial, but the result itself is not. The evolutionary psychologist Irwin Silverman has argued that this sex difference evolved because men used survey knowledge to find their way home after tracking game across the landscape, but women relied more on remembering places where plants or small creatures could be gathered.[24] You, the reader, should be waving a metaphorical red flag at this point. Neandertals had allocentric perception, but not the division of labor envisioned by Silverman. Of course, Neandertals weren't our ancestors, and Silverman's hypothesis might still hold for us. But it does underline how the actual evidence for evolution can complicate hypotheses based solely on modern experiments.

Memory

Something odd has happened to memory, at least to the idea and the meaning of the word. Computer science borrowed it decades ago to refer to data storage and retrieval, and then jazzed it up with bits, bytes, and RAM. Now when scholars refer to memory everyone thinks of the same thing: data retrieval. The metaphor has become the meaning. Human memory is not the same as data retrieval. For one thing, it is not accurate. Whenever we generate a memory it risks being a bit different from the time before. This is why eyewitness testimony is so unreliable. Memory is also not predictably fast; some memories come quickly, but others linger just out of reach on the tip of our tongue, and yet others we know that we know, but nothing we do can elicit the memory. There are also several different kinds of memory, and from a cognitive and neurological perspective they are quite different from one another.

An important distinction in memory is between short-term or "working" memory and long-term memory. Here short is very short, only a matter of a few seconds. The classic example of short-term remembering is a phone number from the point you look it up in a book to the point you enter it into a phone. If you don't keep repeating it to yourself, you lose it. In a sense this is not really memory, but information held in attention. We will treat it separately in the section on planning. Any information retrieval after a longer interval (minutes to days to years) is long-term memory. Here there is a second distinction between semantic and procedural memories. Semantic memories are memories of facts that in theory could be expressed in words, for example, "Abraham Lincoln and Charles Darwin were born on the same day, February 12, 1809." The memory trace itself may not exist in that form, but the information is amenable to verbal expression. Procedural memories are "how to" memories: how to tighten a bolt, how to play an F major scale on a bassoon, how to drive a car. Such memories are much more difficult to express as words but are just as important, perhaps even more important, to the everyday negotiation of our lives. Neandertals undoubtedly relied on both, as all primates do, but the nature and balance of the two may have been a bit different from ours. There is a third kind of long-term memory that is sometimes subsumed under semantic memories: episodic memories. More than just memories of facts, episodic memories recall specific events in one's past. They are sometimes referred to as autobiographical memories. They consist of facts and specific images, but also an awareness of one's position in time (autonoesis) and a reconstitution of the emotional feeling of the event.[25] Episodic memories are difficult to study in the psychology laboratory, but they are very important in human affairs, and their development may have had important consequences in human evolution.

Neandertals clearly possessed semantic memories. We will postpone for the moment a discussion of words (we think that they had them) and focus on factual memory. Neandertals clearly had memories such as "Rhinoceros are prey" and "Rabbits are prey but hardly worth the effort" and "There is a cliff at La Cotte." In fact we suspect that Neandertals had a huge body of such knowledge about their local territories, knowledge that each Neandertal acquired while growing up and journeying with his or her parents and groups. It is possible that most of this was learned firsthand by observation and experience, rather than secondhand through words. The evidence that Neandertal children also led lives of exertion suggests that they traveled with parents as soon as they could walk. There were no Neandertal playgroups or schools. But this does not mean that this knowledge was written on a blank slate of Neandertal neurons.

Humans, and by extension probably Neandertals, come endowed with certain category biases in the way we organize perceptual information. An old bias that probably goes back to early primates is the bias toward recognizing and distinguishing faces. We see faces in pretty much anything that vaguely resembles a face (e.g., the Madonna in a piece of burnt toast), and we begin doing this very soon after birth. Our brains come prewired to do this (in the inferior temporal lobes in a location known as the fusiform gyrus). But faces aren't all. Our brains also come prewired to distinguish animate objects from inanimate objects. We just naturally perceive animals as a special category of thing; we do not need to learn this. And, more controversially, all humans appear to think *about* animals (and plants) by categorizing them into taxonomic groups. In a taxonomy an item belongs to a series of more inclusive groups, such as {[(lion) – carnivore] – animal}. People don't all use the same taxonomy, but they all use some form of taxonomy. Our brains appear to supply the basic structure for thinking about animals, but learning fills in the boxes. We suspect that this particular piece of cognition evolved long ago, and evidence suggests that Neandertals may have relied on it heavily. Almost everywhere that we have good evidence for Neandertal hunting, Neandertal groups zeroed in on one or two varieties of ungulates, at least whenever possible. They didn't just focus on size; they focused on a small range of species. They were clearly parsing the ungulate world in a familiar way: {[(tur) – edible ungulate] – animal} or {[(gazelle) – edible ungulate] – animal}. One of the notable things about Neandertals is how committed they appear to have been to these categories, an allegiance beyond what we might expect from modern hunters.[26]

We have abundant evidence for Neandertal procedural memory. Our best examples come from technology, and we will discuss them at length in Chapter 3. But much of Neandertal hunting and gathering would also tap procedural memories: how to thrust a spear and butcher a carcass; how to transport portions of carcasses; how to follow paths. Neandertals' intimate knowledge of their locales was held in semantic memory, but knowledge of how to make use of terrain and exploit resources would have had a strong procedural component.

Episodic memories are harder to detect. After all, these are memories of events, often autobiographical in nature. It seems to us that the Neandertals who drove the mammoth over the cliff at La Cotte would have remembered this event, probably for their entire lives. It was undoubtedly dramatic and charged with excitement and even fear—just the kind of event that is easily and quickly filed away in episodic memory. But we have no evidence that they did. Later in the archaeological record of *Homo sapiens* there are painted depictions of what appear to be events, but Neandertals never produced anything like this.

Planning

"Planning" is an unfortunately vague term, and its inherent imprecision has led to a contentious and ultimately futile argument in paleoanthropology about Neandertal planning. Consider the following scenarios:

1. Four male chimpanzees sitting on the forest floor hear a lone colobus monkey in the tree above them. They exchange glances, after which three of them climb into adjacent trees, while the fourth ascends the tree with the ill-fated monkey. They chase it, surround it, kill it, and eat it.
2. Four Inuit set out on a day-long sled journey to a pass in the mountains, where they wait. Two days later a caribou herd appears and begins to file through the pass. The Inuit kill many of them.

3. Toward the end of the rainy season, four Native Australians set fire to an area of woodland in order to encourage a second "green-up" of grasses. This in turn attracts kangaroos, which are an important component of the Australians' diet. The next year they set a different tract of land alight. They do not burn the original again for ten or twelve years.

These are all examples of planning, but they are not all the same, in terms either of the effort required or the "futureness" of the tasks. What do we mean when we say "plan ahead"? Certainly the chimpanzees in example 1 were planning ahead, and yet their plan was very different from the plan used by the Australian hunters. Neandertals also planned ahead; no one disputes this. But as a conclusion it just doesn't tell us very much about Neandertals or about planning. We need to break planning down into components; once we do this it will be possible to characterize Neandertal planning in more detail and compare it to what modern humans do.

For any action to count as planned, it must be based on anticipation of a specific future state, and the anticipation must govern a choice of actions. If a lion wakes in the late afternoon, ventures forth and stalks and kills a zebra, is this planning? We don't think so. The lion's action can be more easily understood as the result of hunger and well-learned sets of responses; there is no reason to suppose that the lion anticipates a specific killing scenario. What about the chimpanzees in example 1? Perhaps groups of male chimpanzees always climb adjacent trees when they hear colobus monkeys, in which case we could invoke a scenario like the lion's. But there seems to be more going on than this. The chimpanzees anticipate the specific escape options for the monkey, and though it is perhaps possible that this is just a well-learned routine, it seems more likely that the chimpanzees were "planning ahead," fitting their actions to the anticipated future state. How difficult is this in a cognitive sense? One requirement for successful planning is autonoesis, a subjective sense of time with a past and a future and an ability to place oneself in the past and future. And to place oneself one needs a concept of self as an independent agent. Chimpanzees do have a concept of self—most can learn to recognize themselves in a mirror—and would also appear to have some subjective feel for past and future. It is fair to conclude that they have some degree of autonoesis. But their anticipation of a future state would not have to be nearly as detailed as that of the Inuit or Australian hunters, so there must be additional cognitive components.

One possibility is that the human hunters have many more, and more elaborate, long-term memories of hunting scenarios. Through semantic and procedural memory they just have a much larger range of solutions for any anticipated scenario. And they almost certainly have episodic memories of past hunts. By projecting such episodic memories into the future they would have a ready-made plan of action. We think that Neandertals did a lot of this kind of thing. What worked in the past should work again.

Another difference between chimpanzees and human planning lies in communication. Chimpanzees do cooperate when they hunt, and do appear to communicate a joint goal through glances—but after that they do not send information to one another to help coordinate action. They vocalize their emotional state (e.g., excitement), and other chimpanzees can locate them that way, but they don't vocalize (or use silent signs) to coordinate action. Humans clearly do, and we think that the Neandertals did as well. Effective hunting on the scale practiced by Neandertals would almost require some coordination via communication of information. For this reason (and others) we suspect that Neandertals had language.

Are autonoesis, long-term memory, and communication of tactical information all that is required for planning? There is something still missing, and as a way to understand differences in planning ability, it may be the most important. It is the amount of relevant information an individual can hold in attention and think about at the same time. Psychologists refer to this ability as "working memory," though "working attention" might be a better term.[27,28] To give you a better idea of what working memory entails, here is a classic working memory problem: While reading this paragraph, remember the next to the last word in each sentence, in order, and at the end recite them all back without referring to the text. It is a bit of a pain to do, primarily because you must really concentrate—that is, devote much of your attention to the task. The number of sentences you can do this for is one measure of your working memory capacity. Note that you must do several things at once. First, you must read. Second, you must identify the next to the last word in each sentence. Third, you must remember the words. We would bet that those of you who actually took on the challenge had to repeat the sequence of words to yourself as you tried to read. This rehearsal mechanism is one of the several components of working memory.

In general, working memory is associated with the highest levels of cognition. One's working memory capacity correlates highly with such prized abilities as reading comprehension and scores on tests of general intelligence. Working memory is not simply the amount of information you can hold in attention. It also includes an "executive" component that helps us "think about" what we are attending to. Two of these executive functions that are important to planning are response inhibition and filtering out distraction. Your procedural memories, and evolution, supply you with a huge array of automatic responses to situations confronted in daily life. The fight-or-flight response is a good example. But sometimes you need to suppress such automatic responses. It would be hard to kill a rhinoceros if you fled in fear at the mere sight of it. The executive component of working memory is what you use to suppress that response. Distractibility is also a serious limitation to planning and executing plans of action. (Most readers can attest to this.) People with high working memory capacity are better able to block out distractions and focus on the task at hand. When combined with autonoesis, these executive functions enable us to make more elaborate plans for future action. Refer back for a moment to our examples of hunting plans. They all differ from one another in their demands on working memory capacity. The chimpanzees must suppress their natural response, which would be to chase the monkey immediately, but the Australians must do much more. Not only do they need to inhibit their fear of fire, but they must go to considerable effort and spend a lot of time to organize their actions in anticipation of a reward that will arrive weeks or months later, while at the same time considering the consequences of their choice for ensuing years. In such situations it is very easy for short-term goals or desires to distract from the successful completion of long-term plans.

Our primary evidence for Neandertal planning comes from what we can reconstruct about their hunting. Clearly Neandertals anticipated future events and made decisions with this knowledge in mind. Setting up ambushes on permanent game trails in the Caucasus is conclusive evidence that Neandertals did not wander aimlessly around their territories. But it is also clear that their temporal reach was not as great as that of many modern hunter-gatherers. We have no evidence for changes in tactics within seasons, something very common with modern foragers, and certainly no evidence for anticipating years in the future, as we have for modern Australian hunter-gatherers. Neandertal tactics were very effective and part of a well-learned strategy. But we have no evidence for long-term forecasting or the kinds of response inhibition used by modern hunter-gatherers and agriculturalists to plan for the future. So the evidence as it currently stands points to small but real differences between Neandertals and modern humans in planning ability.

IN SUMMARY, NEANDERTALS were the top predator on the landscape wherever they lived. They focused their hunting on a narrow range of large mammals and shifted down the food pyramid only if pressed. In some regions, such as northern Europe, they hunted mammoth and rhinoceros, and were thus perhaps the only prehistoric people ever to have focused on these dangerous beasts. But Neandertals were adaptable and able to switch focus to whatever large mammals a region had to offer. Their tactics relied on an intimate knowledge of local landscapes, an ability to surprise and ambush prey, and the raw courage to kill them at close range using thrusting spears. They hunted in small groups that included women and children. There was no serious division of labor, except with regard to which hunting task someone performed. Even though they focused on only a few species, they did not travel long distances in pursuit. They hunted in relatively small territories, rarely over 1,000 square kilometers (380 square miles), often just up and down a major river valley. They rarely ventured outside of their home territories. They knew every boulder, cliff, and defile, and used them to advantage. But their focus on a few species and relatively small territories meant that their hunts often failed and they went hungry for days. When they did have a successful kill, they had to gorge themselves in order to average the 3,500 to 5,500 calories per day necessary for their survival. By our standards, theirs was a tough, brutal way of life.

Neandertal hunting required certain ways of thinking and favored certain personality types. It relied very heavily on long-term memory for locations of good hunting; the kinds of animals available; places of hiding, cliffs and defiles, and sources of raw material; and so on. They needed to be able to find their way from any point in their territory to any other, and they used route following and probably survey knowledge to do it. They used short-range planning, at least over the duration of hunting forays. They were tolerant of pain and able to persevere in situations that would cow even the best of us. They were likely taciturn and not easy to excite. And they were very successful, conquering a variety of habitats in northwestern Eurasia, including the harsh glacial climates of western Europe. Neandertal hunters had an impressive cognitive profile, but it was different from that of modern hunter-gatherers.

An important component to Neandertal success at hunting was their use of technology, especially stone-tipped spears. As it turns out, archaeologists know a lot about this topic. It is the focus of the next chapter.

Further Reading

1. Gamble, C. 1999. *The Palaeolithic Societies of Europe*. Cambridge, U.K.: Cambridge University Press.
2. Mellars, P. 1996. *The Neanderthal Legacy: An Archaeological Perspective from Western Europe*. Princeton, N.J.: Princeton University Press.

3. Stiner, M. 1994. *Honor among Theives: A Zooarchaeological Study of Neandertal Ecology.* Princeton, N.J.: Princeton University Press.

4. Callow, P., and J. M. Cornford, eds. 1986. *La Cotte de St. Brelade 1961–1978: Excavations by C. B. M. McBurney.* Norwich, U.K.: Geo Books.

5. Aiello, C. L., and P. Wheeler. 2003. Neanderthal thermoregulation and the glacial climate. In *Neanderthals and Modern Humans in the European Landscape during the Last Glaciation: Archaeological Results of the Stage 3 Project.*, eds. T. van Andel and W. Davies. Cambridge, U.K.: McDonald Institute for Archaeological Research.

6. Binford, L. R. 2001. *Constructing Frames of Reference: An Analytical Method for Archaeological Theory Building Using Hunter-Gatherer and Environmental Data Sets.* Berkeley: University of California Press.

7. Finlayson, C. 2009. *Neanderthals and Modern Humans: An Ecological and Evolutionary Perspective.* New York: Cambridge University Press.

8. Hardy, B. 2010. Climatic variability and plant food distribution in Pleistocene Europe: Implications for Neanderthal diet and subsistence. *Quaternary Science Reviews* 29:662–679.

9. Henry, A., A. Brooks, and D. Piperno. 2011. Microfossils in calculus demonstrate consumption of plants and cooked foods in Neanderthal diets (Shanidar III, Iraq; Spy I and II, Belgium). *Proceedings of the National Academy of Sciences* 108(2): 486–491.

10. Bocherens, H., et al. 2001. New isotopic evidence for dietary habits of Neandertals from Belgium. *Journal of Human Evolution* 40:497–505.

11. Richards, M. P., and E. Trinkaus. 2009. Isotopic evidence for the diets of European Neanderthals and early modern humans. *Proceedings of the National Academy of Sciences* 106(38): 16034–16039.

12. Frison, G. 1991. *Prehistoric Hunters of the High Plains.* San Diego: Academic Press.

13. Gaudzinski, S., and W. Roebroeks. 2000. Adults only: Reindeer hunting at the Middle Palaeolithic site Salzgitter Lebenstedt, northern Germany. *Journal of Human Evolution* 38:497–521.

14. Adler, D. S., et al. 2006. Ahead of the game: Middle and Upper Palaeolithic hunting behaviors in the Southern Caucasus. *Current Anthropology* 44(4): 89–118.

15. Bar Yosef, O., et al. 1992. The excavations in Kebara Cave, Mt. Carmel. *Current Anthropology* 33(5): 497–550.

16. Geneste, J.-M. 1988. Les industries de lat Grotte Vaufrey: Technologie du debitage, economie et circulation de la matiere primiere lithique. In *La Grotte Vaufrey a Cenac et Saint-Julien (Dordogne), Paleoenvironments, chronologie et activites humaines*, ed. J. Rigaud. Paris: Memoires de la Societe Prehistorique Francaise.

17. Hoffecker, J. 2002. *Desolate Landscapes: Ice-Age Settlement in Eastern Europe.* New Brunswick, N.J.: Rutgers University Press.

18. Kuhn, S., and M. Stiner. 2006. What's a mother to do? The division of labor among Neandertals and modern humans in Eurasia. *Current Anthropology* 47(6): 953–980.

19. Goodall, J. 1986. *The Chimpanzees of Gombe*. Cambridge, Mass.: Belknap Press.

20. Lee, R., and R. Daly, eds. 1999. *The Cambridge Encyclopedia of Hunters and Gatherers*. Cambridge, U.K.: Cambridge University Press.

21. Istomin, K., and M. Dwyer. 2009. Finding the way: A critical discussion of anthropological theories of human spatial orientation. *Current Anthropology* 50(1): 29–49.

22. Ingold, T. 2000. *The Perception of the Environment: Essays on Livelihood, Dwelling and Skill*. London: Routledge.

23. Gladwin, T. 1970. *East Is a Big Bird*. Cambridge, Mass.: Harvard University Press.

24. Silverman, I., et al. 2000. Evolved mechanisms underlying wayfinding: Further studies on the hunter-gatherer theory of spatial sex differences. *Evolution and Human Behavior* 21:201–213.

25. Tulving, E. 2002. Episodic memory: From mind to brain. *Annual Riview of Psychology* 53:1–25.

26. Atran, S. 1990. *Cognitive Foundations of Natural History: Towards an Anthropology of Science*. Paris: Cambridge University Press.

27. Baddeley, A. 2007. *Working Memory, Thought, and Action*. Oxford: Oxford University Press.

28. Coolidge, F. L., and T. Wynn. 2005. Working memory, its executive functions, and the emergence of modern thinking. *Cambridge Archaeological Journal* 15(1): 5–26.

3

Zen and the Art of Spear Making

HOW DO YOU make a stone-tipped spear? What do you need to know, and what skills must you possess? On the surface, it seems pretty simple, but is it actually easy to do? When we were children we both played at cowboys and Indians (it was a politically incorrect age) and in the course of our pretend play tried to make bows and arrows. Among our many technical failures, the most maddening was our inability to attach our triangular rocks to our arrows. String wouldn't hold them, at least not for long, nor would glue, nor would wire; the force produced by even our pathetic bows was just too strong. It had seemed like a simple task, at least to a ten year old, but it wasn't. It is actually a minor engineering feat, combining materials with different physical properties in such a way that force is passed between them linearly with a minimum of torque. Neandertals solved the engineering design problem of attaching points to shafts, and the result was arguably the first true invention in technological evolution, a weapon that not only enhanced their ability to kill large animals, but enhanced their ability to harm one another, as the Shanidar #3 skeleton attests. So what did Neandertals need to know in order to make a stone-tipped spear?

Spears

Neandertals needed spears, or more specifically the idea of a spear: an object that extended one's reach and could penetrate the skin of an animal. This bit of Stone Age technology is actually much older than Neandertals, and may even go back to some of our earliest ancestors. Understanding the uses of a sharp stick is well within the abilities of chimpanzees, who occasionally use them to stab and extract bush babies (small nocturnal prosimians) from their sleeping nests. Chimpanzees do not sharpen the sticks to increase their effectiveness, but our early ancestors might have, in particular early members of our genus, *Homo erectus*. The archaeologist Manuel Dominguez-Rodrigo has analyzed the edges of stone tools from the site of Peninj in Tanzania (about 1.4 million years old) and discovered a residue of acacia wood. *Homo erectus* had used the tools to scrape or cut wood, and Dominguez-Rodrigo suspects that they had been making spears.[1] He did not find actual spears, but this is not a surprise. Spears are made of wood, and wood rarely preserves in archaeological deposits for more than a few hundred years. There are a few exceptions to this rule, from sites where artifacts have been preserved by bogs or marshes (where oxygen, the major culprit in decay, doesn't circulate). One of the most spectacular of these sites is Schöningen.

Schöningen is a coal mine, a huge open-cast mine near Hanover in Germany from which large pieces of heavy equipment extract coal. As part of the coal company's environmental policy, archaeologists examine all areas that will be added to the mine, and in the early 1990s these archaeologists struck archaeological gold: the remains of three spears found with the skeletons of horses. Later they found four more spears. These spears are almost 400,000 years old and were made by the ancestors of Neandertals, *Homo heidelbergensis*.[2,3] The spears resemble nothing so much as giant toothpicks. They are about 2 meters (6 feet) long and pointed at both ends. The *heidelbergensis* artisans who made them had cut down young spruce trees, removed the side branches, and shaved off the bark and outer layers of wood, all with stone tools. They then shaved the spears so that they had a center of gravity about one-third of the way from one end, a weighting that resembles that of modern javelins. Hartmut Thieme, the archaeologist in charge, thinks that the spears were thrown at the horses.

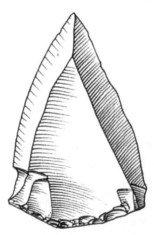

Some archaeologists doubt this; after all, how much damage could a giant toothpick inflict on a horse, even if thrown by a strong individual? Enough to bring it down? But few doubt that the spears were used to stab and ultimately kill the horses, which the hunters then butchered.

Schöningen is one of those rare sites that not only is intrinsically interesting, but also solves a long-standing puzzle. Archaeologists had found many sites of the same period that yielded stone tools and animal bones, but these associations could have resulted from something simpler and less dangerous than spear hunting, such as scavenging. The Schöningen spears effectively put this minimalist explanation to rest: Neandertals'

FIGURE 3-1 A Levallois point (after Mellars[6]).

ancestors had hunted large animals using spears. It was clearly an established form of hunting, and technology, as Neandertals evolved. At some point later in time, perhaps as early as 200,000 years ago, Neandertals escalated the pre-historic arms race by attaching a sharp stone point to the ends of their spears.

In Neandertal times the simplest (not the easiest!) spear point was a flat, triangular-shaped stone tool that was sharp along two edges, something archaeologists call a "Levallois point" (Figure 3-1). They were effective, but they were very difficult to make. If you want to think like a Neandertal, you must learn to think in stone.

Stone Knapping 101

In his groundbreaking television series *Cosmos*, the late Carl Sagan rhapso-dized about a universe filled with billions and billions of stars and galaxies, an immense and virtually endless source of data for astronomers. Archaeologists studying human evolution have a more modest vision. We have literally millions and millions of stone tools—so many, in fact, that simple counting is some-times a problem. In the 1970s the archaeologist Michael Mehlman excavated a site on the Serengeti Plain in Tanzania with an eye to studying the tools from the Middle Stone Age (roughly 200,000 to 40,000 years old, contemporary with Neandertals). But by the time he had dug down to the MSA levels, he had already recovered more than 200,000 stone tools from more recent periods. As much as he wanted to ignore them, archaeological ethics wouldn't allow it. He needed to examine each one of those 200,000, at least in a cursory manner, and report the results. The task took years to complete. Stone tools are actually a bit like galaxies. At first sight they all seem pretty much the same. But the longer you look and the closer you look, the more differences you see, and the more patterns you can identify, all of which call out for explanation. For over 99.7 percent of human evolution stone tools were the mainstays of technology.[4,5] They accompanied our ancestors from the time we were just bipedal apes in tropical Africa 2.6 million years ago to the invention of metallurgy six thousand years ago. And they changed over that vast period of time, changes that mark important evolutionary developments in diet, social interaction, and cognition. And, of course, they were essential to Neandertals' way of life.[6,7]

The basic idea behind stone tools is deceptively simple: breaking stone produces sharp edges. Chimpanzees and orangutans (and college students) have been taught to do it, so this basic idea did not require some great intel-lectual leap on the part of our ancestors. It certainly wasn't rocket science, at least not at first. The most basic procedure in stone knapping is hard-hammer percussion. ("Knapping" is the formal term for making stone tools.) The knapper holds a stone in one hand (the left hand for a right-hander) and strikes its edge with another, harder stone held in the other hand. The receiving stone is called a "core," and the striking stone a "hammer." If the blow is powerful enough and lands close to an edge of the core, it will break off a piece, which is called a "flake." Flakes can have very sharp edges, in some cases sharper than a surgeon's scalpel. Not all kinds of stone are good for knapping. Soft, grainy stone like sandstone won't hold an edge, and hard, crystalline rocks like granite are too hard and have a crystalline structure that hinders knapping. From the time of the earliest known stone tools, from a place called Gona in Ethiopia, knappers selected the best available local raw material, even if it was not the most common.

Even the simplest stone tools are useful for a variety of tasks. At first our bipedal ape ancestors used them for just two things: sharp flakes for cutting and cores for bashing. We know that they used the flakes to cut meat from scavenged carcasses, and the cores to smash into bones for marrow (cores have sharp ridges marking the places flakes had been detached), because archaeologists such as Selesi Semaw, who excavated Gona, have found the bone fragments along side the stone tools.[8] We suspect they cut and pounded other things as well, but the products haven't preserved for archaeologists to find. For about the first one million years of technical history this is about all we see—hammers, cores, and flakes—used primarily to process scavenged carcasses and perhaps other food. But about a million and a half years ago we begin to see some technical changes. The subsequent history of stone tool technology is dominated by two themes: modifying the edges of flakes into greater and greater varieties and managing the volumes of cores so that they supply more or better flakes.

The edges of a freshly knapped flake can be very sharp, especially if produced on high-quality stone such as obsidian or flint. Sometimes, however, a razor-sharp edge is not the best tool for a job. Let's say you want to scrape fat off an animal hide. A flake with a razor-sharp edge will slice into or even through the hide unless it is used carefully and slowly. Also, sharp edges are thin and break easily. But you could make the edge a bit sturdier and a bit less sharp by removing a series of tiny flakes from the edge. Archaeologists term such edges "scrapers," based on their presumed function. Scrapers began to appear in archaeological sites about 1.8 million years ago. Other varieties of edge appeared about the same time, including flakes with projections that could have been used to punch holes. For the next 1.3 million years stone knappers produced more and more varieties of flake edges, so that by 200,000 years ago Neandertals were producing scores of slightly different edges on flakes.

The second trend is more interesting from a cognitive point of view. Stone knappers developed ways to make more effective use of cores. If you just bash on a core to knock off flakes, the core very quickly becomes more or less spherical, with no acute edges that can be broken to produce flakes. This is what chimpanzees do when trained to knap. If you examine the core carefully and look for the best places to strike, you can keep the core productive longer. This is what the earliest hominin stone knappers did. And if you want the core to be even more productive, you devote some effort, and flaking blows, to controlling the overall shape of the core as it dwindles in size. Such techniques are known as "core preparation" techniques, and they didn't begin to appear until about half a million years ago. But why bother preparing a core? Why not just pick up another stone and start knapping anew? The answer is that high-quality stone is rarely at your fingertips, or even close by, when you need the tools. It often has to be carried from its source, and stone is heavy. It is more effective to make efficient use of the material you have than to go get more, a trip that could take hours, or even days.

Stone tools were not like Phillips-head screwdrivers; with very few exceptions that appeared very late in prehistory, stone tools were not so tied to single functions. Sharp flakes could perform a variety of tasks, and they could be easily modified to perform a different task. A single flake rarely required much effort to make (ignoring for the moment core preparation) and was not a prized possession. Most were discarded soon after use. More maddening for archaeologists, a knapper who lived thousands of years later could pick up an old tool, put a new edge on it, and be in business. Hominins probably didn't carry many sharp flakes around; it was simpler to carry a core and maybe a hammer and produce flakes when necessary. This is a different relationship with tools than the one we have with ours; if we wanted to tune up a 2005 Volkswagen but didn't have the very specialized tools necessary, we would be out of luck.

Stone Knapping 499

We are now in a position to revisit spear points. Just how do you knap a thin, flat, sharp flake with sides that taper to a point? Neandertals used two rather different strategies. In one they modified a large flake by removing small flakes from the sides until they had achieved the appropriate pointed shape. In the second they prepared a core in a way that would allow them to knock off a triangular flake that would need no further modification. As the latter is the more interesting, we will focus on it.

The first step was to acquire some decent stone, and this was sometimes a problem for Neandertals. They did not like to venture very far. As we saw in Chapter 2, Neandertals had intimate knowledge of their local territories and used established routes and paths to encounter game, but they rarely if ever traveled beyond the boundaries of their well-known places. The same held for raw material. They knew where to find the highest quality stone in their own territory. But, as with animal carcasses, carrying was a limiting factor; no one wanted to carry heavy quantities of stone very far. As far as we can determine from archaeology, they almost always used local stone, available within a few kilometers of where they knapped it. Once in a long while they carried very high quality stone up to 250 kilometers, but this was rare. And in these cases they never wasted any of it; archaeologists find only small pieces discarded after repeated use. They reserved their most sophisticated techniques for high-quality stone, much like a modern cabinetmaker devotes his or her best efforts to mahogany or cherry but lavishes less attention on pine. And the most impressive knapping technique of the time was Levallois.[9,10]

Levallois is a core preparation technique in which a knapper controls the volume and shape of a core in order to have greater control over the size

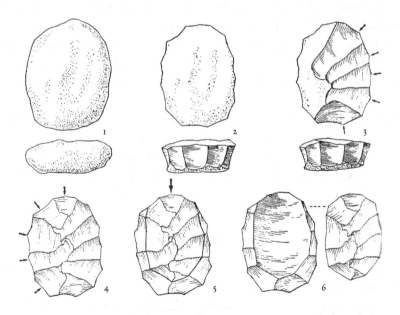

FIGURE 3-2 Steps in the basic Levallois core preparation technique (after Mellars[6]). The arrow in view 5 indicates the striking platform, the point where the final blow must fall to strip off the flake in 6. The goal of Levallois was to produce one or more large, thin flakes that could then be either used as is or modified further.

and shape of a few flakes. In basic Levallois the knapper first removes flakes from a mass of stone (usually flint) to produce what is sometimes called a "tortoise" core, because it resembles an inverted tortoise shell (Figure 3-2). If done well, the knapper can strip off most of the "top" of the core (the bottom of the tortoise shell) as a single large, flat flake. The knapper could then trim the oval flake into the shape of a point. But some Neandertals used an even more clever form of Levallois. They prepared the upper surface of the core in such a way that the resulting flake was triangular in shape, with two converging sharp edges, a kind of natural spear point (Figure 3-1). It required no further modification before it could be mounted on the end of a spear.

Hafting

The first engineering design in the history of technology may have been the hafting of a stone point onto a wooden shaft. Think about it for a second. This was not a new knapping technique that someone could have stumbled upon by accident, or by modifying another technique ever so slightly. This was a novel idea. Someone had to decide to attach a stone tool to a wooden shaft, and then experiment with ways to do it. He or she had no precedents to work from; nothing like it had ever been done before. We call it an engineering design because there were unforgiving laws of physics to overcome. Force applied to the point must be channeled linearly to the shaft with a minimum of torque. All lateral forces must be restrained somehow, or the point will break off of the shaft. And if you are stabbing a mammoth, there will be a lot of force. With the sharpened spruce spears of Schöningen this was not an issue; the spear was a single piece. But they weren't very sharp, or nearly as deadly as a spear mounted with a sharp, stone point.

So how did Neandertals do it? Archaeologists have direct evidence of one of the solutions. From a site in Syria they have excavated a point whose base was covered with bitumen, a kind of natural asphalt.[11] These Neandertals glued the point on. This is not as simple as it sounds. First, they had to acquire the bitumen, then they probably had to heat it to make it workable, after which they had to position the point on the shaft in such a way that the bitumen only channeled the force to the shaft; it would not have been strong enough to bear the direct force itself. In western Europe Neandertals used pitch made from birch bark.[12] Other Neandertals may have tried tying the point onto the shaft, but there are no direct remains of such lashing. Archaeologists have found triangular flakes that were broken in a pattern suggesting that they

FIGURE 3-3 A Neandertal spear with a point hafted on the end, in this case by lashing.

had been mounted on shafts, but they lack an adhesive.[13] These points may have been lashed on (Figure 3-3).

However they did it, Neandertals solved an important engineering problem, thereby enhancing the effectiveness of their primary weapon.

Innovation

Hafting was an impressive bit of problem solving, but it appears not to have been typical. It was a new solution to an old problem—how to make spears more deadly and reliable—but such innovation was extremely rare in Neandertal technology. In fact for the entire time that Neandertals lived in Europe and western Asia, some 200,000 years, they produced only a few technical innovations. Among stone tools it is possible to identify perhaps a score of innovations over the entire period, and all of these were modest variations on previous procedures (e.g., producing triangular flakes using Levallois technique). Neandertals made the same range of tools for the entire period. The percentages of each variety fluctuate over time and space, but the plain truth of the matter is that 40,000-year-old tools are almost indistinguishable from 200,000-year-old tools. We have no reason to suppose that archaeologically invisible components of technology (fiber, clothing, etc.) fared any differently. This dearth of innovation is one of the clear differences between Neandertal and modern technology. It suggests that something was very different about either Neandertal culture or the Neandertal mind.

Thinking through Iron and Stone

Earlier we suggested that if you want to understand Neandertals you need to understand how to "think through stone." What could this possibly mean? It sounds a bit like New Age crystal gazing rather than scientific archaeology and psychology. But we do have a serious goal—to capture something of the nature of technical thinking—and traditional laboratory-based experiments miss something important when it comes to the real-world experience of tool use. Here is an example of what we mean. On a recent TV program on barbarian weapons presented by the History Channel, a blacksmith demonstrated one of the steps in making a Celtic sword. At one point he said, "You must be in the fire with the sword." This is certainly an odd statement. He didn't climb into the forge, or even peer closely at it. So what was his point? In his own fashion he was expressing something actually rather important: for an experienced artisan, tools are extensions of perception, and hence extensions of mind. This may strike many readers as fanciful, or at best a curious metaphor, but this is because most educated people, in the West at least, have acquired a "dualistic" understanding of the mind and body. In essence this dualistic notion holds that minds and bodies are separate. Thinking is what minds do; the brain is a physical thing, but it is the organ of thoughts, which reside wholly in the mind. The body is a servant of the mind, and not involved with thinking. With such a dualistic understanding, it is difficult to describe what the smith was getting it, or indeed much of what we actually do in everyday life. A more familiar example may help. Experienced drivers learn to feel the road through the steering mechanism of the automobile. If they are driving in snowy weather and hit a patch of ice, they feel the traction change and slow down or take other action before they have conscious awareness ("Oh, s***!"). They have extended their perception and awareness through the mechanical linkages of a machine. Such decisions bypass conscious attention; the body reacts to lost traction before conscious realization of the hazard. In a case like this, it is misleading to conclude that everything of interest takes place in the head. The brain controls some of this response, but peripheral nerves and "trained" muscle groups are also important. Recently psychologists have begun to take such phenomena seriously, and a perspective known as "embodied cognition" has arisen to investigate them. Psychologists have not yet devoted much attention to tool use, but anthropologists have.

Charles and Janet Keller have written an account of how a modern blacksmith thinks while working iron.[14] It is an account that captures much of the feel of technical thinking and also describes some of the important organizational features of tool use. Any technical task is accomplished through a sequence of steps; this much is common sense. Just think of the steps in baking a cake or assembling a bicycle. (Did you read the instructions?) It is how an experienced artisan conceives of the steps that is most interesting. According to the Kellers, the blacksmith first comes up with a general plan, a "mental outline," if you will, of an entire task based on time available, materials and tools at hand, past experience, and of course an understanding of the goal and some idea of the steps required. The smith then assembles tools and materials based on this mental outline and begins. Each step is a largely self-contained segment of activity focused on an immediate goal (e.g., twisting an iron rod for a railing). It is here that technical thinking becomes a bit harder to render into words. Much of the thinking within a step is done through materials and tools. Much of the knowledge of how to complete the step emerges from the tools available, the material to be worked, and the immediate goals. Tools and materials literally trigger images, but not necessarily words or sentences. These images are not simple visual representations, but complex associations of sound, vision, postures, and muscle tensions. And they are not static mental pictures, but instead are dynamic, continuous actions. After completion of a step, the smith goes on to the next step, and so on until completion. So a technical procedure consists of relatively discrete segments or bundles of activity performed sequentially. In retrospect it is often possible to describe the activity as a hierarchy of routines and subroutines, but this is not how the smith thinks while working, and it is certainly not the way the activity was learned. Any hierarchy is an artifact of our analysis. John Gatewood, an anthropologist who worked on a salmon fishing boat, described the technical procedures as strings of beads, and this is an apt metaphor for technical thinking.[15]

The interaction between tools, materials, and the artisan produces flexible responses. Technical thinking is not rote repetition of a memorized sequence. The smith continuously monitors the state of tools and materials, a monitoring that is done by feel, by sound, and by sight, and based on this information adjusts the procedure in subtle or not so subtle ways. He or she does not have to reinvent the wheel. Instead the smith, or any artisan, taps into a huge body of alternatives learned over the course of his or her technical life. Each time an artisan completes a task the result will be at least slightly different because of differences in materials and the immediate conditions of the task that will dictate slightly different responses. When artisans repeat the same task over and over, more and more of the decisions become routine, and even responses to typical problems become more or less automatic. But this level of ability takes time to acquire.

Apprenticeship

How did Antonio Stradivari learn to make a violin? How did your grandfather learn to tune an internal combustion engine? Neither read a book. Indeed neither had much verbal instruction at all. They learned by watching and doing, trial and failure, with perhaps a kind, or more likely a rude, word here and there from a mentor. As a young man Stradivari spent several years in the workshop of a master luthier, Nicolo Amati. In your grandfather's case he probably learned from his father, or perhaps an older brother, a more informal arrangement but similar to Stradivari's in most major respects. Apprenticeship is still the way people learn to use tools. An acquaintance of one of us traveled to Japan in the 1970s to apprentice to a cabinetmaker as part of her dissertation research. To her great disappointment, she discovered that apprentices were not even allowed to touch tools for more than a year. She was allowed to sweep up and watch. This initial drudgery is in fact quite typical for apprenticeships and has often appeared in literary accounts. (Remember the hapless Mickey Mouse in *Fantasia*'s version of "The Sorcerer's Apprentice.") Moreover it was, and is, not unusual for apprenticeships to last up to ten years—more than enough time to earn a PhD.

We do not learn language by apprenticeship. In fact we receive no training in language at all. Young children acquire language if they are raised in an environment with language, even if it is impoverished in a linguistic sense. There are things that parents do that help, but they are not necessary. This difference in the way we learn tool use and the way we learn language suggests that there is something very different about these two varieties of thinking. Tool use relies on many kinds of knowledge, few of which can be captured adequately by the written or spoken word. But there is something even more fundamental: tool use and tool making rely on cognitive abilities that are only superficially similar to language and that evolved earlier than language. When we learn by apprenticeship we tap into this older ability. It is an ability that is also a major component of many of our most revered activities: music, sports, visual arts, and dance.

Expertise

Experimental psychologists may not have paid a great deal of attention to practical technical thinking, but they have studied two kinds of activity that, we believe, tap into the same cognitive abilities. These activities are sports and expert performance.

High-level performance in sports has long been a fascination for not just participants and fans, but also physiologists and psychologists. Most readers will be vaguely familiar with sport physiology, if through no other avenue than well-publicized doping scandals. Sport psychology is probably less familiar but is equally important to the competitor. Both of us have firsthand experience in amateur competitive sports—one in tennis, the other in fencing—and have some firsthand knowledge of what it takes to be successful, as probably do many of our readers. What does it take, psychologically, to be a champion, or even be just good? Reading and speaking are not high on the list of necessary abilities; how-to books just do not present much that is helpful, and professional athletes are notoriously inarticulate. Coaching can help, but how? Just what does a good coach do? It is actually far from clear what makes for successful coaching. Yes, coaches can get you to practice, which is very important, and also motivate you to do well (by fear or affection), but they cannot really do anything that directly affects your motor abilities; you must do this yourself. Practicing is essential for success in any sport. Ask any professional or accomplished amateur athlete, and he or she will report having practiced for thousands of hours and perhaps tens of thousands of repetitions. Why? If you know how to do it, why can't you just do it? The answer is that motor activities tap into a kind of thinking known as "motor memory," which is a variety of procedural memory. You can teach your motor memory only by, literally, teaching the muscle groups, the nerves that run them, and brain regions that control them. The only way to do this is by repetition; the system is almost completely cut off from conscious control. So how do you know what pattern to repeat? Someone might, we suppose, explain it to you in words (but try describing a tennis serve using only words), or someone could show you, which is much more effective. Humans (actually all primates) have motor neurons that fire when an action is observed. They are called mirror neurons, and they help us learn motor activities by copying. But we still have to repeat the pattern—again and again and again—before we get good at it.

We think that it will not be hard for the reader to see that much of technical thinking—pounding a nail, hammering a Celtic sword, knapping stone—relies on motor memory. And like the athlete, the artisan requires thousands of hours and tens of thousands of repetitions in order to achieve proficiency. Equally, it is difficult for the artisan to access the knowledge consciously. He or she often relies on visualizing techniques to guide performance. (Remember the smith being "in the fire" with his Celtic sword?) But the reader will likely have detected significant differences between athletes and artisans as well. The segmented, string-of-beads organization used by the artisan would be unusual for an athlete (except perhaps divers or gymnasts). To understand this a bit better we need to examine expert performance in other activities.

Experts are individuals who have achieved a high level of performance in a narrow field of activity.[16] In psychology the most studied experts are mnemonists, people who can rapidly memorize long strings of digits, sometimes hundreds of digits in length, and chess masters, the people who can play multiple games blindfolded and win every one. These are not autistic savants, who have remarkable memory or calculating abilities but are otherwise poor performers in the everyday social world. Experts are normal people who have trained in their area of expertise. We will use chess as an example. What, exactly, is the chess master doing as he or she plays multiple games blindfolded?

1. He or she assesses the problem—the board position—very rapidly. How? Partly by using a cognitive shorthand. Standard chess openings have names. If the expert just remembers that game #1 started with "Sicilian defense," he or she will instantly know the position of all thirty-two pieces on sixty-four squares. From then on it is a matter of remembering one move at a time. In other words, the expert taps into long-term memory and activates patterns that are stored there.
2. He or she is able to respond quickly and can switch very rapidly to alternative solutions when necessary.
3. His or her responses are almost error-free. There is a very high degree of accuracy and reliability.
4. He or she could be distracted from the task and return to it without losing task-relevant information. Somehow he or she can restore relevant information without loss. Think of memorizing a sequence of digits, such as a telephone number. For most of us any interruption, say a conversation with a spouse, would cause us to forget the sequence. An expert mnemonist would lose nothing. Similarly our chess expert can switch attention to another game and then return to the first game without loss of information.
5. But our chess expert's ability is narrowly restricted to chess. He or she would not be better at Go or checkers than anyone else.
6. Our chess expert needed years to attain this proficiency. The number most cited in the literature on expertise is ten years, that is, ten years from novice to mastery. This number applies in all areas of expertise, not just chess. How long did it take Tiger Woods to develop his skill at golf, or Anne-Sophie Mutter her skill at the violin? Sure, there are prodigies, but even they require years of practice to reach their zenith.

How do experts do it? What kinds of thinking do they use? Is their kind of thinking unusual, found in only the few truly talented individuals, or is it more mundane and familiar? It is, in fact, a common form of thinking that all of us use. Everything an expert knows is stored in the neural network of long-term memory (LTM). Psychology is a bit vague about what qualifies as long, but in general it is any period of time longer than a few seconds. Most of the information supplied to us by our senses is lost after a few seconds; there is rarely a reason to store it. Sometimes, if sensory experience occurs at the same time as an emotionally charged event, we file it into LTM easily. (Where were you when you heard of the 9/11 attack?) But most of the time if we want information to go into LTM we need to repeat it several times.

Experts practice for years in order to build up huge libraries of information and procedural skills tied to their field of expertise. They rely heavily on two techniques, sometimes called "chunking" and "chaining." Chunking divides up large bodies of information into smaller, more easily processed chunks, and chaining links the chunks together by using an end bit of information in one chunk to stimulate the recall of the next. Let's say you need to memorize a piece of music. How do you do it? You could play it or sing it hundreds of times until you've got it, or you could break it down into smaller pieces, memorize one phrase, then the next, and so on, using the ends of the phrases to prime the next phrase. A similar approach works in sport. A fencer learns a new attack by breaking it down into elements, learning each element, chaining them together into a sequence, and then repeating them again and again and again, until they flow without conscious attention. Action in a fencing bout goes much too fast to "think" through an action—too fast for an attentive decision—except at the highest strategic level. It is the high number of repetitions in practice that produces the high reliability of expert thinking. The links in the chain are so strong that once initiated they rarely fail. As the expert acquires more and more chunks strung into longer and longer chains, he or she begins to organize them into routines and subroutines, and thus begins to build a hierarchy of information. This too facilitates memory.

But isn't long-term memory generally unreliable? And isn't it slow? When asked for the dates that Millard Fillmore served as President of the United States, how many of us have the answer at our fingertips? We might be able to remember, but it will probably take a few seconds for us to search our memory. Yet expertise is not slow. The information is available instantly. How?

Experts rely on shortcuts. For example, the chess master attaches a label (the psychological term is "cue") to all of the information about a particular opening (Sicilian defense), and just by recalling the cue he or she accesses all of the information held in LTM and does not have to search for it laboriously. Similarly in sports, a particular posture or action by an opponent will trigger a response from motor memory that is deployed instantly. But like the long-term memories themselves, the expert must practice for years to assemble an appropriate body of cues and link them to appropriate responses held in LTM.

Expertise is best exemplified by experts in the fields of art, sports, and craftsmanship, but it is a style of thinking that we all use, everyday. A popular, some say overworked, example is driving to work. Most of us do this every workday and think little of it. It is in fact an expert motor performance. You must direct the action of a machine that weighs a ton or more, traveling at 100 kilometers per hour (60 miles per hour), which will respond to even small errors in guidance by crashing into a tree or another vehicle. Most of us become so adept at doing this that the cues and procedures operate with little conscious attention. Most readers will have had the experience of arriving at work or home, at some point realizing that they have no memory of the trip, and wondering how they avoided being killed. Expert thinking can operate without conscious monitoring by using visual or sound cues that access procedural and motor responses. It is no less impressive for being automatic; that is the reason for its success. And it isn't just driving: cooking, gardening, even video gaming are expert domains. Indeed the U.S. Air Force has found that some of the expertise acquired in video gaming transfers easily to guiding remotely operated drone aircraft. (Don't tell your adolescent children.) Expertise is not limited to experts. It is ubiquitous in modern thinking. Experts have just honed their abilities and take expertise to its extreme.

Technical thinking is expert thinking. The procedures and techniques deployed by the blacksmith are held in long-term memory and learned through chunking and chaining. The smith uses cues to call up responses appropriate to the array of problems encountered for any task. It is these cues that allow the rapid switching to alternative procedures and that supply the flexibility necessary for effective craftsmanship. Perhaps the clearest indication that technical thinking is expert thinking lies in the way it is learned: by apprenticeship. Just as it takes years of practice to attain chess mastery or a professional proficiency at the violin, it takes years of practice to become a smith, or a cabinetmaker, or a master stone knapper.

Neandertal Technical Thinking

We have lingered over our discussion of technical thinking for two reasons. First, it is an understudied and underappreciated component of the modern mind that deserves wider understanding; second, it is the component of Neandertal cognition that we can see most clearly. Archaeologists have literally millions of Neandertal stone tools. We know how Neandertals made them, what they used them for, and where they found their raw material. In some cases we can trace actual sequences of action from the point a Neandertal began to knap a core to the point he or she quit. We have enough information to compare Neandertal technical performance to that of modern artisans and ask if it was organized in the same way, if it was embodied in the same way, and if it was learned in the same way. And we will see that the similarities outweigh the differences.

Marjorie's Core

Once in a while archaeologists are privileged to excavate a site where the artifacts and spatial patterns of artifacts have been little disturbed by natural processes over the intervening years. This is very unusual for sites that are tens of thousands or hundreds of thousands of years old; there has just been too much opportunity for mechanical and chemical and biological processes to alter and move the remains. But sometimes archaeologists get lucky. One such site was Schöningen; another is Maastricht-Belvedere in the Netherlands, excavated by Wil Roebroeks and others from Leiden University.[17] They found nothing quite as spectacular as the Schöningen spears, but they did excavate the remains of several episodes of stone knapping. Because it was an open site—that is, not in a cave—the episodes were separated in space rather than on top of one another, and because natural processes had moved the artifacts very little, the archaeologists were able to "refit" some of the cores. Refitting is basically assembling a three-dimensional jigsaw puzzle: the archaeologist takes all of the flakes that had been knapped from a core (identified largely by color and texture of the stone, in this case flint) and reassembles them into the original core. It is rare to be able to refit more than ten flakes in this way, which results in only a partial core, but at Maastricht-Belvedere archaeologists (Marjorie among them) were able to refit thirty-eight flakes into a classic Levallois core. Because the refit core was almost complete, archaeologist Nathan Schlanger was able to describe a long sequence of action, identifying in the process many of the decisions made by the Neandertal knapper.[18]

The knapper first prepared the core by trimming flakes off the sides, top, and bottom to produce the classic tortoise shape (Figure 3-4). He or she did this initial preparation somewhere else; none of the initial flakes was found at Maastricht-Belvedere. In the first phase that Schlanger could document, the knapper initially searched the top of the core for a potential "distal convexity." This is the place on the curved upper surface of the core that will be the end of the Levallois flake. The knapper then prepared a striking platform

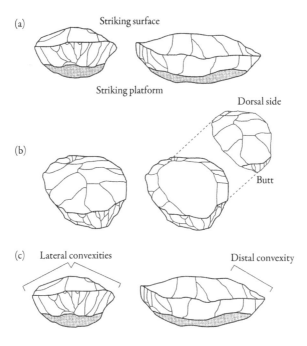

FIGURE 3-4 A more nuanced view of a Levallois core, identifying some of the important features (and visual cues) that governed the size and shape of the final flake (after Chazan[10]). The distal convexity was an especially important cue for the knapper of Marjorie's core.

on the opposite side of the core. This platform is the place the knapper will strike to remove the Levallois flake. It is essential that the angle between the platform and the top of the core be within a narrow range or the flake will not "propagate" correctly; the blow will not peel off the top of the core. Schlanger found that the knapper completed this first phase by striking off a single, large Levallois flake that in fact did take off much of the top of the core. He or she then rotated the core laterally 90 degrees and examined the recently flaked upper surface for a new usable distal convexity (Figure 3-5).

Because he or she had rotated the core, the new distal convexity was a lateral convexity in the previous phase. He or she then removed three small flakes, running across the core top, that produced a new set of lateral convexities, in effect restoring the classic tortoise shape. Next came preparation of a new striking platform, followed by a major blow to remove a Levallois flake. This time, however, the knapper was not as successful and did not remove most of the core top. He or she was able to strike a second flake from the same

FIGURE 3-5 Plan and profile views, and plan drawing of Marjorie's core (after Roebroeks[17]).

platform. He or she then rotated the core 90 degrees and began anew. Schlanger was able to document seven such phases. Of course, the core became smaller and smaller with each phase, but Schlanger noticed something very interesting: the Levallois flakes themselves did not get smaller and smaller. The knapper was able to adjust his or her technique to maintain flake size.

Marjorie's core reveals a number of important things about what the knapper was thinking.[19] First, the action was divided into discrete phases, or steps. Second, each phase was guided by a distinct perceptual cue, the distal convexity, that guided the action within each phase. Third, the knapper responded to changing conditions of the core, adjusting technique to maximize Levallois flake size and maintain core productivity. Fourth, there was an overarching hierarchy to the entire task, with an overall goal (Levallois flakes), subroutines (the phases), and sub-subroutines (identify distal convexity, configure lateral convexity, prepare platform, strike off Levallois flake). This was not ad hoc flaking or a rote sequence. It was a flexible strategy of core reduction, very much like the Kellers' mental outline. Finally, the knapper followed at least one rule: rotate the core 90 degrees before examining it for a new distal convexity.

Marjorie's core is a clear example of expertise. The knapper responded to changing conditions by subtly changing his or her approach, and did this by responding to perceptual cues that called up appropriate responses from long-term memory. What Marjorie's core tells us is that Neandertal technical thinking was no different in its basic organization than modern technical thinking.

Hafting

Archaeologists are not able to describe the process of hafting in the same detail as they can describe stone knapping because most of the components haven't survived. Our primary source of insight comes from trying to duplicate the process ourselves. Lyn Wadley has done this for South African Middle Stone Age spears.[20] The spear makers heated acacia gum, combined it with powdered ochre (a mineral pigment about which we will have much to say in a later chapter), and perhaps honeycomb, and used this glue to attach the stone barbs to shafts. There was clearly an understanding of fire as a transforming agent. And the process took several days, with many discrete steps, a clear example of expertise. But these MSA spear makers were not Neandertals. They were in fact modern humans. Neandertal artifacts have been associated with pitch made from birch bark, and such pitch cannot be found naturally (unlike bitumen). Thus we know that Neandertals were practicing a bit of chemistry in preparing glue. Unfortunately no one has yet done an experimental investigation comparable to Wadley's for these Neandertal glues. On the surface it would appear comparable, at least in the necessity of using fire. Even if it wasn't as complex as the MSA hafting, it was still a technological achievement that relied on combining a stone tool and a spear by means of a haft. But how did they come up with the idea in the first place?[12,21]

Innovation

Where do new ideas come from? Is necessity the mother invention, or is this just an adage of our engineering age? Did some Neandertal Thomas Edison devote 1 percent inspiration and 99 percent perspiration to inventing the spear haft? Does it matter? Readers might well argue that we will never know how it happened, and so it is pointless to speculate. And they would be at least half right. We will never know the details, but we can combine what we know about modern innovation with some of the things we know about Neandertal technology and come up with a likely scenario. We do think that innovation matters, and that the nature of Neandertal innovation is an important clue to their thinking. But we do need to be careful or we will build a circular argument; if we argue that Neandertals innovated by combining previous technical systems into new solutions because this is typical of expert systems, we cannot then use this form of innovation as evidence for expert cognition! Our evidence for expertise must come from elsewhere, as it has.

We will follow a fairly conservative dictionary definition of "innovation": the introduction of new methods or procedures, by whatever means. There is a bizarrely contentious debate in some corners of academia (mostly schools of business…) about definitions of innovation and invention, which we will avoid. We will reserve "invention" for innovations that are "thought out" in some sense. It is a narrower term with implications of attention and design not carried by the broader concept of innovation.

Many technical innovations occur through daily use of tools. Serendipitously a stone knapper may strike a core in the wrong way and get an unanticipated but useful result. If he or she recognizes the result as useful (a big if), can duplicate the mistake (another if), adds the technique to his or her repertoire, and others copy it (another big if), then innovation will have occurred. We imagine that many, perhaps most of the innovations in stone knapping over the 2.6 million–year Stone Age occurred just this way. A second form of innovation occurs through tinkering with the techniques one has learned. If our stone knapper just messes about one day, trying out odd ways of holding a core or different material for a hammer, and an interesting result occurs, this too may lead to innovation, with the same "if" caveats as earlier. Neither serendipity nor tinkering qualifies as invention, and neither is able to change technology in dramatic ways. From an archaeological perspective we would see modifications of existing technologies but nothing entirely new. However, the *rate* of change produced by serendipity and tinkering might be significantly different, given that one is passive and the other active. Both would affect components of expert cognition, but tinkering just might require greater tolerance for alternatives in the procedural memory schemes. Those using rigid procedures, no matter how well learned, are unlikely to tinker.

Invention differs from innovation in requiring some kind of "thinking out." Quite simply, the inventor has to think about it first, which is not required by serendipity or tinkering. There are many ways to invent. A simple way is to combine two previously separate techniques to solve a novel problem. A more sophisticated way is to devise general principles of force, resistance, and so on—that is, a kind of physics—and use them to design a new tool.

Could hafting have emerged serendipitously or from tinkering? Hafting a spear point just might have emerged through tinkering if there had been a previous technology of hafting something else, but this simply shifts the question to the something else. Hafting seems to require invention; some Neandertal thought it up. We suspect, in the absence of evidence, that Neandertals may have had a technology of gluing; that is, they may have used natural bitumen or made birch pitch to stick things together. Or they may have had a technology of tying, perhaps bundles of wood or carcass parts. And then some Neandertal added one of these technologies to their thrusting spears. Such an invention would require an ability to hold all of the relevant information in attention and manipulate it. In cognitive science this is known as working memory. All mammals have working memory, but the amount of information that can be processed varies considerably. Invention requires more working memory capacity than tinkering because separate technical procedures have to be coordinated, and the only way this happens is through working memory. Thus the active invention of something like hafting suggests a larger working memory capacity than was typical of earlier hominins such as *Homo heidelbergensis*.

There is one very important piece of evidence about which we must remind the reader before we equate our Neandertal with Werner von Braun: Neandertals almost never came up with new ways of doing things. As important as it is to understand how Neandertals might have innovated, it is important also to remember that they almost never did. And this virtual absence is perhaps the single most important difference between Neandertal technical thinking and ours.

Learning and Language

Neandertals almost certainly learned to make and use tools in much the same way that we do: by observing someone else, copying the procedure, and trying again after failure. Neandertals wouldn't have had formal apprenticeships; their social groups were too small and unspecialized to need them. Moreover all Neandertals knew how to knap. Some may have been more adept than others, but as we will see in the next chapter, it is unlikely that Neandertal children had more than a few active adults from whom to learn, and shopping for a good tutor was not really an option. A Neandertal child almost certainly learned from watching a close relative.

The cognitive context of technical learning is almost entirely embodied; that is, hands, muscles, and peripheral nerves are important components. This is true for us and was true for Neandertals. By repeating actions over and over again the novice Neandertal built up longer and longer procedural chains and, equally important, extended his or her perception to encompass more and more of the task in time and space. This extension included sight, the ability to detect and focus on relevant information in the visual field; hearing, assessing a core by how it sounds when struck; touch, attending to the texture of raw material; and motor sensation, recognizing how a muscle group feels when delivering a blow with the right force. Little of this information exists as words. Indeed Neandertals, like modern artisans, would have had a hard time describing these actions in speech. Novice Neandertals did not need to sit through long, boring lectures on the theory of Levallois knapping. They learned it by doing it.

But passive observation may not have been quite enough for the most difficult procedures; some active instruction may have been necessary. For example, the importance of the distal convexity as a focal point in Levallois technique may not have been obvious to the casual observer. Someone would have needed to draw the observer's attention to that particular feature. This is shared attention, and it does not require speech. At the risk of boring the reader, another fencing example will illustrate the potential of speechless instruction. Fencing masters occasionally use a technique for advanced students known as *leçon muet*. The maestro explains nothing. Instead, while in fencing position, the maestro will present his or her blade in a specific way (usually a sequence of moves) to draw the pupil's response. If the pupil responds appropriately, the maestro may acknowledge with a nod; if not, the maestro will present again (or, if particularly annoyed, smack the pupil on the mask with the side of his or her blade!). The pupil must concentrate on all of the subtleties of the maestro's action in order to learn the new response. An entire lesson can go by with no words spoken. In bypassing the verbal step (and verbal thinking), *leçon muet* delivers instruction where it matters: to motor memory and situational response. Of course, stone knapping is not fencing, but it too is a complex motor task with subtle changes in situational cues. It is necessary to draw a novice's attention to certain cues, but this could be done in any number of ways: pointing, repetition, vocalizing positive and negative sounds. Sure, a few words here and there might facilitate the exchange, *but this is not how the important information is conveyed*.

Shared attention relies on an ability cognitive scientists call "Theory of Mind," the ability to imagine what another sees and knows. A minimal kind of Theory of Mind is gaze following, the ability to follow another's gaze to see what he or she is looking at. A more advanced form of Theory of Mind is the ability to conceive of several levels of intentionality: I know that Fred knows that I know that he is currently in Spain enjoying himself while I write this paragraph. One of the more interesting trends in the evolution of primate cognition is the trend in Theory of Mind. Chimpanzees, for example, learn to recognize themselves in a mirror (the concept of self is a low level of intentionality), but gorillas apparently cannot. Humans are generally able to handle four or five levels of intentionality, such as the Spain example above.

Shared attention is also a matter of Theory of Mind and levels of intentionality. The instructor relies on Theory of Mind, and so does the novice. In observational learning the novice who can imagine what the master sees and knows will have an advantage over a novice who cannot. With instruction through shared attention, the instructor must be able to imagine what the novice sees and understands in order to direct the novice's action. The best instructors will also be able to imagine what the novice imagines about the instructor's own understanding, one additional level of intentionality. This is certainly true of modern instruction, but what about Neandertals? It is just possible that Neandertal instruction could have worked effectively without the addition of the final level of intentionality; that is, Neandertal instructors may not have needed to imagine what the novice knew about the instructor's own knowledge. But it certainly would have facilitated learning complex techniques such as Levallois.

TO SUMMARIZE, TECHNOLOGY was absolutely essential to Neandertals' survival. Their success and well-being depended on their ability to make and use tools. Not surprisingly, they were very good at it. We see this most clearly in their stone knapping, where they mastered and perfected one of the most difficult stone-working techniques: Levallois. But they were not slaves to a single way of doing things. They varied their knapping techniques according to the nature and availability of raw material and regularly squeezed maximum use out of their tools by modifying them into several successive tool types.

In this chapter we have emphasized stone knapping because it is the technical domain about which paleoanthropologists have the most information. It is reasonable, we think, to conclude that other aspects of Neandertal technology must have used similar processes. We have also introduced readers to several important ways to understand the kind of thinking that goes on in tool use: embodied cognition, expertise (motor and procedural memory, chunking and chaining, etc.), apprenticeship, shared attention (and Theory of Mind), and working memory. Today we continue to rely on these kinds of thinking more than most people appreciate. They are the basis for many of our most valued accomplishments in sports, craft, art, and music. Just as important, each of us uses them every day when we drive to work, cook a meal, mow a lawn, or play a video game. And in doing these things, we are thinking like Neandertals.

The only major difference between Neandertals' approach to technology and ours lies in the area of innovation. Modern human technology is characterized by rapid, cumulative technological change, a phenomenon that depends upon innovation. Neandertals almost never innovated. Indeed there is only one clear example of Neandertal invention, hafting, though we suspect that there must have been others yet to be identified. Neandertals could and occasionally did invent novel technical procedures, but they did so only rarely. They did a lot of tweaking, making slight changes to established techniques, but active invention was just not a common component of their technology. For tens of thousands of years they were content with their basic technological profile. Why? Their slightly lower working memory capacity might have been one factor, but we doubt that this could have been the only factor.

Cumulative technological change also requires social transfer of knowledge and expertise. New information must spread if it is to become established, and this requires an effective social network. As it turns out, Neandertal social networks may have been quite different from ours and those of our recent ancestors. Neandertal social behavior and organization are the topics of Chapter 4.

Further Reading

1. Dominguez-Rodrigo, M., et al. 2001. Woodworking activities by early humans: A plant residue analysis on Acheulean stone tools from Peninj (Tanzania). *Journal of Human Evolution* 40:289–299.

2. Thieme, H. 2005. The Lower Palaeolithic art of hunting: The case of Schoningen 13 II-4, Lower Saxony, Germany. In *The Hominid Individual in Context: Archaeological Investigations of Lower and Middle Palaeolithic Landscapes, Locales, and Artefacts*, eds. C. Gamble and M. Porr. New York: Routledge.

3. Haidle, M. 2009. How to think a simple spear. In *Cognitive Archaeology and Human Evolution*, eds. S. De Baune, F. L. Coolidge, and T. Wynn. Cambridge, U.K.: Cambridge University Press.

4. Bordes, F. 1968. *The Old Stone Age*. New York: World University Library.

5. Gamble, C. 1999. *The Palaeolithic Societies of Europe*. Cambridge, U.K.: Cambridge University Press.

6. Mellars, P. 1996. *The Neanderthal Legacy: An Archaeological Perspective from Western Europe*. Princeton, N.J.: Princeton University Press.

7. Dibble, H. 1987. Reduction sequences in the manufacture of Mousterian implements in France. In *The Pleistocene Old World*, ed. O. Soffer. New York: Plenum.

8. Semaw, S., et al. 2003. 2.6-million-year-old stone tools and associated bones from OGS-6 and OGS-7, Gona, Afar, Ethiopia. *Journal of Human Evolution* 45:169–177.

9. Boeda, E. 1994. *Le Concept Levallois: Variabilite des Methodes*. Paris: CNRS Editions.

10. Chazan, M. 1997. Review of *Le Concept Levallois: Variabilite des Methodes,* by E. Boeda; *The Levallois Reduction Strategy* by P. van Peer; and *The Definition and Interpretation of Levallois Technology,* edited by H. Dibble and O. Bar Yosef. *Journal of Human Evolution* 33:719–735.

11. Boeda, E., J. Connan, and S. Muhesen. 1998. Bitumen as hafting material on Middle Palaeolithic artifacts from the El Kowm basin, Syria. In *Neandertals and Modern Humans in Western Asia*, eds. T. Azakawa, K. Aoki, and O. Bar-Yosef. New York: Plenum.

12. Grunberg, J. 2002. Middle Palaeolithic birch-bark pitch. *Antiquity* 76(1): 15–16.

13. Shea, J. 1993. Lithic use-wear evidence for hunting by Neandertals and early modern humans from the Levantine Mousterian. In *Hunting and Animal Exploitation in the Later Palaeolithic and Mesolithic of Eurasia*, eds. G. Peterkin, H. Bricker, and P. Mellars. Washington, D.C.: American Anthropological Association.

14. Keller, C., and J. Keller. 1996. *Cognition and Tool Use: The Blacksmith at Work*. Cambridge, U.K.: Cambridge University Press.

15. Gatewood, J. 1985. Actions speak louder than words. In *Directions in Cognitive Anthropology*, ed. J. Dougherty. Urbana: University of Illinois Press.

16. Ericsson, K. A., and W. Kintsch. 1995. Long-term working memory. *Psychological Review* 102(2): 211–245.

17. Roebroeks, W. 1988. *From Find Scatters to Early Hominid Behaviour: A Study of Middle Palaeolithic Riverside Settlements at Maastricht-Belvedere (The Netherlands)*. Analecta Praehistorica Leidensia vol. 21. Leiden: University Press of Leiden.

18. Schlanger, N. 1996. Understanding Levallois: Lithic technology and cognitive archaeology. *Cambridge Archaeological Journal* 6(2): 231–254.

19. Wynn, T., and F. L. Coolidge. 2010. How Levallois reduction is similar to, and not similar to, playing chess. In *Stone Tools and the Evolution of Human Cognition*, eds. A. Nowell and I. Davidson. Boulder: University of Colorado Press.

20. Wadley, L., T. Hodgskiss, and M. Grant. 2009. Hafting tools with compound adhesives in the Middle Stone Age, South Africa: Implications for complex cognition. *Proceedings of the National Academy of Sciences* 106:9590–9594.

21. Pawlik, A. and J. Thissen. 2011. Hafted armatures and multi-component tool design at the Micoquian site of Inden-Altdorf, Germany. *Journal of Archaeological Science* 38: 1699–1708.

4

A Focus on Family

THE OLD MAN of La Chapelle-aux-Saints, a Neandertal we introduced in Chapter 1, was really not so old. Current estimates place his age at a little over thirty at the time of his death, barely old enough to have finished graduate school. Anthropologists of a hundred years ago misinterpreted his physical condition—tooth loss, arthritis, etc.—as reflecting an advanced age. But as we saw in Chapter 1, most of the wear and tear on Neandertal skeletons resulted from the rigors of their way of life. "It's not the years, it's the mileage," or so the adage goes. La Chapelle must have had a difficult time negotiating Neandertal daily life. For one thing, he could not have chewed his food. Neandertal diets in western Europe at the time of La Chapelle were almost entirely carnivorous, and most animal products require at least some chewing. The wear on Neandertal incisors, and scratches across the incisors of many Neandertals, suggest that Neandertals may have habitually held large pieces of meat with their front teeth while they cut off a chewable chunk with a stone knife. La Chapelle might have been able to do this, but he could not then have chewed the chunk. Perhaps he diced the meat into small pieces, but they would have had to be very small. Perhaps someone did the chewing for him, a solution often practiced by modern people with the same problem. He also would not have been much help on the hunt. He could not have been fast, nor delivered powerful blows, nor carried much. Yet he had clearly lived for several years with these handicaps. He needed help, and someone provided it. And when he died, someone took the time and effort to scoop out a shallow depression in a cave floor, place him in it, and cover him over.[1–4]

Neandertal life was social life. Like modern people and all primates, Neandertals lived out their lives in social groups. Also like modern people and other primates, much of their way of thinking evolved to handle the challenges of social life. Reconstructing Neandertal social life from the bones and refuse that they left behind is more difficult than reconstructing their diet or technology, but it is not impossible. Neandertals left many clues, and many puzzles, that allow us to contrast their social life—and cognition—with ours.

Family

Neandertals spent their daily lives in very small face-to-face groups, which are groups of individuals in more or less direct contact, if not physically, then within eyesight or hearing. Field studies of nonhuman primates have revealed a large array of solutions to the problem of how many individual primates can congregate at one time, and the factors controlling face-to-face group size are both ecological and reproductive.[5] Male orangutans spend most of their time alone, while savanna baboons often live and travel in large troops of more than a hundred individuals of all ages and sexes. Modern human hunter-gatherers, however, are not so diverse in terms of group size. A hunter-gatherer group is typically made up of a number of families, consisting of a man, a woman, their offspring, and often parents or siblings. The families are usually related to one another by blood or marriage (or in a pinch, "fictive" kinship, as when in Western culture we refer to a close family friend by the term "aunt" or "uncle") and also usually have social ties to individuals in other face-to-face groups. The total number of individuals varies from perhaps thirty for a small face-to-face group to two hundred for a very large one. Often hunter-gatherer face-to-face groups will join together for part of the year with related groups, a pattern determined largely by ecological factors.[6] We emphasize simple size of groups here because that is a social factor that we can detect in the archaeological record. So what about Neandertals?

Lazaret Cave is located near Nice on the French Riviera and was used by Neandertals during a cooler interval toward the beginning of the next-to-last glacial period. The cave itself is located about 100 meters above and 500 meters back from the coast of the Mediterranean. It is a long, narrow cave running into the mountainside, with an area of slightly over 300 square meters. The Neandertal occupation area covered only about 35 square meters (about 350 square feet) near the mouth of the cave. Archaeologists directed by Henri de Lumley carefully excavated this living area, recording the precise location of every bone, stone flake, and bit of charcoal, exposing a pattern of debris that reflects Neandertal activities in the cave.[7] As is apparent in Figure 4-1, the Neandertals invested time and energy in rearranging rocks on the

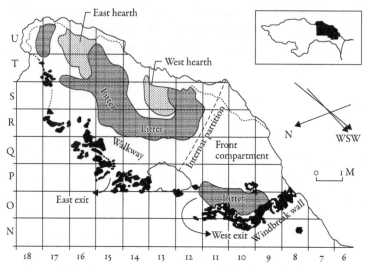

FIGURE 4-1 The floor pattern at Lazaret cave (after de Lumley[7]).

cave floor. They moved large rocks from the living area and piled them along the periphery. De Lumley has argued that these rocks must have anchored some kind of wall or windbreak, but many archaeologists are skeptical because there are no postholes, discolored columns of earth that mark the locations of decayed wooden posts. What is clear is that Neandertals did improve their living space, if only to remove inconvenient rocks. They also built two fires against the cave wall (more on the fires later). Stone knapping debris is littered throughout, as are broken-up animal bones, but the most curious finds are the remains of small marine mollusks. They were too small to have been a food source, so what were they doing in a cave 100 meters above the beach? De Lumley suggests, reasonably, that they were carried in on seaweed that Neandertals used for bedding, and he notes the presence of fox and wolf foot bones as evidence of the hides used to cover the bedding.

The clearest implication deriving from the remains at Lazaret is that this Neandertal face-to-face group was very small, probably in the range of five to ten people. If anything, Lazaret is large for a Neandertal cave occupation. The occupation area in the Spanish site of Abric Romani, which we will discuss in Chapter 5, covers about 150 square meters, but many well-documented Neandertal cave sites in Ice Age western Europe are smaller even than the 35 square meters of Lazaret. These were just not large spaces. It is possible, we suppose, that our sample of Neandertal sites is atypical because caves were secondary in importance. But sites in the open tell a similar tale.

Cave sites have natural limits, the walls of the cave, that constrain the extent of the habitation refuse. Often archaeological refuse extends out of the front of the cave or rockshelter, but for Neandertals this extension of the cave area was never very great. Sites in the open have fewer natural limits, and the extent of the refuse better reflects patterns of human use, but natural processes of erosion also have more opportunity to move and alter the archaeological remains. Archaeologists have relatively few well-preserved open sites for Neandertals, but the few we do have present a consistent pattern. Stone tools, debris from stone knapping, and animal bone (if preservation is good) are scattered about the occupation area. There are often patches of ash or charcoal attesting to the use of fire, but rarely anything resembling a constructed fireplace. Most of the time there is little coherent pattern to the refuse. One exception to this is the Ukrainian site of Molodova 1, which was occupied about 45,000 years ago (Figure 4-2).[8] The lateral extent of the archaeological remains at Molodova 1 is about 1,200 square meters, more than thirty times the size of Lazaret. Animal bones, stone tools, mammoth tusks, and ash litter the site. But here there seem to be some patterns: concentrations of stone tools and bone fragments bounded by denser concentrations of large bones and mammoth tusks. One of these enclosures contains the remains of a hearth. Soviet-era archaeologists argued that these were the remains of huts, but like Lazaret, there are no telltale postholes marking the position of posts. Most archaeologists think that the patterns reflect structures of some sort, probably windbreaks. Size-wise, each of the two more obvious structures is a bit larger than the one at Lazaret. If the structures and refuse at Molodova represent a single occupation, then the face-to-face group was several times larger than the one at Lazaret. If each windbreak sheltered five to ten people, and there were four such structures (a generous interpretation), then the face-to-face group included between twenty and forty people. But this is a fairly big "if." The structures could have been made by a single smaller group that returned on several occasions. As the evidence currently stands, we just don't know, though the likelihood of larger groups seems high.

Nothing in the direct archaeological remains of Neandertal sites requires that we posit face-to-face groups larger than the one at Molodova. But there is some provocative indirect evidence: sites such as La Cotte. One mammoth could easily have supported a Molodova-size group for days. Carcasses from eleven could support many more. And recall that all of the animals had been butchered. Perhaps a bonanza such as La Cotte brought together many Neandertal face-to-face groups for the length of time that carcasses were edible.

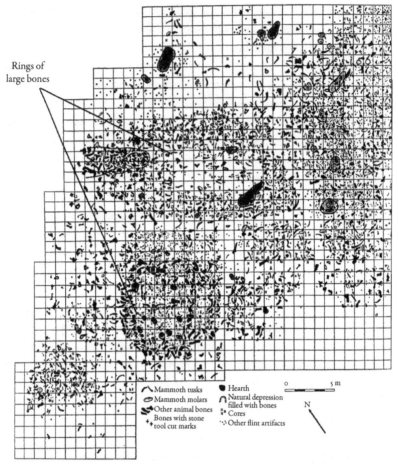

Rings of
large bones

FIGURE 4-2 The distribution of tools, bones, hearths, and possible structures at
Molodova I (after Hoffecker[8]).

In sum, the archaeological evidence indicates that Neandertals spent
most of their lives in small face-to-face groups of as few as five to ten, perhaps coming together occasionally into larger groups of twenty to forty, with
rare aggregations of even more. What social models would fit this evidence?
Neandertals do not quite fit social models of modern hunters and gatherers.
But superficially at least, this pattern does resemble the fission-fusion pattern
of modern chimpanzees,[9] who live in permanent communities of twenty to
eighty organized around related males. These communities almost never congregate in a single assembly; instead they travel and forage in smaller groups
of five to ten whose membership changes often. But there are good reasons to
be cautious in such a comparison.

First, Neandertal territories were much bigger than chimpanzee ranges. Chimpanzee communities typically occupy ranges of 5 to 10 square miles, though in more open habitats they can be considerably larger. Such a range is small enough that the small foraging groups regularly encounter one another. Some chimpanzee vocalizations, especially the food barks, carry for long distances, so that distant foraging parties can congregate at particularly large patches of food when available. Based on the movement of raw material, we know that Neandertal territories were easily ten times as large as those of chimpanzees. Such large territories effectively prohibit the kind of frequent shifts in small party membership that are typical of chimpanzees. And vocal communication could not have carried far enough to keep Neandertal foraging parties in contact; loud vocalizations by hunters at La Cotte would not have attracted a nearby group. We should note here that use of large territories is a very old feature of hominid adaptations that probably goes back to the earliest *Homo*.

Second, Neandertals, like all *Homo*, almost certainly had a social organization based on provisioning. Humans rely on regular, intentional sharing in which foragers acquire and carry food to provision other individuals. Among modern hunter-gatherers both adult women and men bring food back to camp to be distributed to their families and relatives. Humans are unique among primates in this regard. Pair-bonding between adult males and females has evolved, anthropologists believe, in order to maximize the efficiency of provisioning. Men and women provision their mates and offspring, a system that reduces male aggression over access to females (more on this later). Almost everyone agrees that it must have been true of Neandertals, in which case they were very different from chimpanzees, who do not provision and rarely even share.[10]

Third, Neandertal hunting was very different from chimpanzee foraging. Chimpanzees need to forage in small parties because their primary food, ripe fruit, is available only in relatively small patches, too small to support more than five to ten individuals at a time. As we saw in Chapter 2, Neandertals' foraging problems were very different. The patches were often huge, but finding them was difficult. Fig trees do not move around, so chimpanzees can learn and remember their locations; mammoths are a different story. Distribution of food is an important determining factor of not only territory size, but also social organization. Neandertal reliance on small, mobile groups was an efficient means of finding the game they relied on, but was not the most efficient way to consume it. We suspect Neandertals must have had some means of uniting the larger community when possible.

Thus even though the size of the typical face-to-face group was similar for chimpanzees and Neandertals, the underlying ecological and social dynamic was very different.

What was the makeup of a Neandertal group in terms of age and sex? The kinds of people found in a Neandertal face-to-face group were determined partly by economic necessity—finding and processing food—but also by life history and population structure. "Life history" refers to the timing of major developmental milestones, such as weaning, age of sexual maturity, and senes-cence.[11] Such timing is important to evolutionary success because it is directly tied to a species' ability to continue itself through sexual reproduction. For example, if women reach reproductive age later in development, but reach senescence (postreproductive age) sooner, they will have fewer offspring. And if infant mortality is high, there may be too few offspring to maintain a viable population. Changes in life history patterns have been important in primate and human evolution. Humans, for example, have shorter natural birth inter-vals than apes, which means that a human mother can have more pregnancies in her lifetime than an ape mother. But a higher number of offspring puts strong pressure on a woman's ability to provide calories for her offspring, through either nursing or foraging. Humans solved this problem with provisioning; someone, either a mate or a grandmother, helped by bringing the mother food. This particular evolutionary development occurred long before the advent of Neandertals and early modern humans, both of whom relied on provisioning.

Reconstructing life history patterns requires an ability to document age at maturity and other milestones from fossils. The best avenue for doing this is through teeth. It is not only possible to document sequences of erup-tion (e.g., of permanent incisors compared to molars); it is also possible to determine the chronological age of these developments by counting certain growth lines in tooth enamel (perikymata). Neandertals appear to have been very like modern humans in their dental development. There is some evidence that they may have reached dental maturity (eruption of the wisdom teeth) slightly earlier than we do, perhaps by the age of fifteen, but this is within the range of modern variability.[12,13] Paleoanthropologists also have the skeletons of a few very young individuals (including a newborn), and these also appear to fall in the range of human development for brain size and skeletal matu-rity. Thus there is nothing that we know about Neandertal life history that appears very different from ours: they didn't develop much faster, and they didn't reach reproductive age much earlier.

Membership in face-to-face groups was also influenced by popula-tion structure, the relative numbers of men and women, and especially the numbers of old and young. We have no reason to think that the sex ratios of Neandertals at birth were any different from those of modern humans (a slight bias toward boys). Nor do the rigors of Neandertal life appear to have afflicted women more than men, or vice versa. What little we can reconstruct about Neandertal obstetrics suggests that women may have had a *slightly* easier time in childbirth than modern women. Their pelvic opening was a tad bigger, but the neonate's head was not. (But remember, we have only one neonate to work with.) Aside from childbirth, which almost certainly took its toll on Neandertal women, it is probable that men suffered higher mortality rates than women, but this is true of modern humans as well. Where we do find a significant difference is in the number of older people; there were just fewer old Neandertals than we see among modern humans. If we simply divide the adult population into an older group that includes those at least twice as old as first reproductive age (e.g., 30 is twice 15), and a younger group that encompasses those between first reproductive age and twice that, a significant pattern emerges.[14] Only about one-third of adult Neandertals were older when they died, but two-thirds of modern humans were from the older group (based on a skeletal sample of modern humans who lived 20,000 years ago). Given the dangers of Neandertal life this difference is perhaps not surprising, but it does have implications for Neandertal social life. First, many adults died young, at an age when they had young children. Second, there were only a few older adults available, and most of these were probably in a physical condition similar to La Chapelle or Shanidar #1—broken down and marginally productive. In other words, adult knowledge, experience, and labor were in short supply, compared at least to the modern humans who followed them. In this, Neandertals were no different from other human populations living at the same time, including our ancestors.[15] Life in the Palaeolithic was tough all over. But the implications of the dearth of older adults may have had different implications for Neandertals, who, as we discuss below, spent most of their lives in small face-to-face groups.

Men Are from Mars, Women Are from... Mars

When natural scientists compile lists of behaviors that are characteristic of humans and distinguish us from other primates, pair-bonding between adult males and females always makes it toward the top of the list. Pair-bonding is the establishment of a long-term paired association in which the male and female have more or less exclusive sexual access to one another and associate on a daily basis. The pair-bond itself has an emotional component: the individuals establish a strong emotional attachment to their partner. Now humans are not the only primates that establish pair-bonds. The gibbons and siamangs, small apes who live in Southeast Asia, establish lifelong pair-bonds; indeed they are the only real monogamous primates. But the monkeys and our close cousins among the great apes do not establish such bonds; instead they have multiple sexual partners, access being determined largely by dominance relationships.

Pair-bonding is a solution to a perennial problem faced by all sexually reproducing organisms: finding sexual partners. Without sexual partners one cannot reproduce. Natural selection is quite clear on this issue; if you don't reproduce, you don't pass on your genes. Primates have evolved a variety of social solutions to the problem of sexual access. Most often aggression or the threat of aggression (dominance) is a component of mating patterns. But aggression is costly in terms of calories, and selection will opt for less energetically expensive solutions when it can. Pair-bonding does not eliminate aggression, but it does greatly reduce the frequency. This is the reason that male and female gibbons are almost the same size; monogamous mating eliminates selection for large size in the males, who no longer must compete daily for dominance. But pair-bonding is in fact unusual among mammals. It is often a better strategy for males simply to try to mate with as many females as possible, a motivation that leads to intermale aggression. Humans are not immune to such competition among males, so why do we establish pair-bonds? The answer appears to be that human pair-bonding is not just about sex. There is also an important economic component. Males bring food to females and their offspring, thus increasing the female's access to calories and increasing her reproductive success. And if the two have a successful pair-bond, her success would be his success.

When social scientists approach the topic of sexual access and reproduction they eschew the rather analytic "pair-bond" in favor of "marriage."[16] And rather than emphasizing the evolutionary consequences, they focus on the nature of the interaction itself, including its emotional and cognitive components. If sociocultural anthropologists have learned anything about people in the past century and half, it is about marriage. People around the world have developed a variety of marriage patterns, none of which is in any sense typical, let alone correct. But they do have several things in common. First, marriages are contractual arrangements in which the parties agree, often committing themselves legally, to certain rights and obligations. Second, marriage is invariably about children and the potential to have children and who those children "belong to" (are related to). Third, marriage is about labor and property; it is an economic agreement. The partners agree not just to supply sex, but to contribute their labor and wealth to the benefit of the other. For most people today, and for most of recent human evolution, this economic agreement is and has been based on a division of labor. Men supply certain things and perform certain activities, while women supply a different set of things and perform different activities. Note that romantic love, one emotional basis for pair-bonding, is not a necessary component of marriage. Social scientists do not deny its existence or its importance in human affairs (no pun intended); it is just not required for marriage, and indeed can often be a handicap. (Older adults know this, which leads them to arrange marriages for their children, who have not yet figured this out.) Marriage is too important to be left to romantic love.

Neandertals almost certainly had pair-bonds, though paleoanthropologists disagree about when in our past this behavioral characteristic evolved. Some argue that it evolved along with bipedalism over 4 million years ago, and others argue that it didn't evolve until the appearance of early members of the genus *Homo,* about 2 million years ago.[17] The debate centers around energetic requirements of hominin females. Compared to males, females require more calories to keep themselves going and for pregnancy and nursing. As a result females tend to be smaller than males, who need only devote calories to self-maintenance and fighting other males. The "sexual dimorphism" in regard to size is true of all anthropoid primates except the gibbons and siamangs, who are, if you recall, monogamous, with lifelong pair-bonds. Thus if paleoanthropologists could identify a point at which sexual dimorphism was reduced in our ancestors, this would be a provocative clue for the presence of pair-bonding. All paleoanthropologists see evidence for this with *Homo erectus* fossils; the disagreement concerns earlier forms of hominin. Neandertals were about as sexually dimorphic as modern humans, and it is reasonable to conclude that pair-bonding with provisioning was an essential component of their social lives.

But did Neandertals know romantic love? There are both cognitive and physiological components to human sexual attraction. Many are clearly the result of natural selection or, more narrowly, sexual selection. Characteristics that make an individual more attractive to the opposite sex or better able to compete for a mate will become more common over time. But there are also cognitive and physiological traits that predispose you to form and keep a more or less exclusive bond for at least a time. Romantic love would appear to fall into this category. Evolutionary psychology has made a concerted assault on the components of the human mind that have evolved to maximize our ability to find, keep, and even cheat on mates.

In fact romantic love is so unusual, in so many ways, that it takes an almost equally unusual theory to account for it. Evolutionary psychologists begin their reasoning with a major discrepancy: theoretically men can produce millions and millions of children, although the Guinness Book of Records claims that Genghis Khan holds the men's record, with about 10,000 children (and modern genetic studies support this idea), but women are limited by the number of eggs they can produce in their lifetime, about 360 or so. The present record for women is debatable, but it seems to be between 64 and 69 children (with many multiple births). Evolutionary psychologists also reason that given this vast discrepancy in producing children, men's and women's most successful strategies would also vary dramatically. As Genghis Khan figured out, one successful strategy for men for ensuring the survival of their genes is to make love to as many women as possible. In this scenario the man is not helping provide for his children's welfare, and so these children are less likely to survive; however, by having many, many children, there's the likelihood that some will survive to have children themselves, and thus the man's genes survive. Now this would not be a good strategy for women. A woman's best strategy for seeing that her children survive and have children themselves is to take really good care of the few children that she can have. But, evolutionary psychologists reason, she'll need help getting food and other provisions that will ensure the survival of her children, and of course, this is where the man comes into play. If she can get a man to stick around and help feed and provision the children that they have together, her children are much more likely to survive to adulthood. But what does she have to trade that the man wants? Sigmund Freud knew the answer: sex. According to evolutionary psychologists, women trade sex for the provisioning of their children. Why is there red lipstick (and, as we will see, Neandertals did use red ochre for something)? Evolutionary psychologists think that it would be less likely a man would stick around if he knew exactly when a woman could get pregnant. So they believe that women evolved their monthly cycle to make it difficult to know exactly when a woman could get pregnant. After all, a human egg can be fertilized for only about 48 hours out of a month. If a woman's private parts turned bright red only during these 48 hours, as happens in some monkeys, then a man might pay attention to her for only 2 days out of 30, sort of like dogs. Now, women could have developed a strategy in which they just lied about when they could get pregnant; however, lies are far more effective if the liars actually believe them. Thus natural selection came to hide exactly when a woman could get pregnant. She herself doesn't know, so it is unnecessary to lie. So why did humans ever design anything as weird as red lipstick?

According to evolutionary psychologists, it's an unconscious attractant to a man. It reminds him of red, swollen, fertilizable private parts of a woman back million and millions of years ago.

So what's love got to do with it? In economic terms, there are thousands and thousands of men with billions and billions of sperm. This makes sperm cheap. There are thousands and thousands of women but there are many, many fewer eggs, and this makes eggs very expensive. And remember, a woman's best reproductive strategy is to keep that man around to provision her and her child and make sure he doesn't run off after she becomes pregnant. So women can afford to be choosy. And evolutionary psychologists think that women therefore evaluate a man based on whether he'll be a good father. Certainly men also search for good mothers, but they are also looking for fertile young women and plenty of them. So how can a man convince a woman that he will be a good father and stick around, when it's only partly in his best interest to stick around and provision this woman (and his child) when there are so many other fertile women around? Romantic, head-over-heels, crazy love is the answer. Natural selection evolved romantic love so that the man could not only convince the woman that he will love her forever and never stray, but also convince *himself* that he will love her forever and never stray.

Although this reasoning is offensive to some people, evolutionary psychologists mount a large amount of indirect evidence for it. For example, why do surveys of married people always show that men cheat more than women? Because it is consistent with one of men's basic evolutionary reproductive strategies. As we've already noted, it doesn't necessarily help a woman to cheat. Her best reproductive strategy is to make sure her children survive. But a man can bridge both strategies: he can pretend to be a good father and husband, yet still fool around and create a few other children. And please keep in mind that evolutionary psychologists do not offer these hypotheses as justifications for men's behaviors! They are not condoning or pardoning these behaviors, and they are not judging the morality of these behaviors. They are simply trying to explain why women wear red lipstick, why people fall in love, and why people get married and promise to do so until death do them part. So yes, Neandertals undoubtedly fell in love. But whether or not there could have been a Neandertal Romeo and Juliet hinges on other issues as well, including marriage.

If Neandertals had marriage, it was different from marriage as practiced by modern people. The division of labor was not the same. As discussed in Chapter 2, modern hunting and gathering men and women have complementary but distinctly different economic roles. For tropical foragers, the primary economic role of women is gathering plants and small animals, and the primary role of adult men is hunting medium-size to large animals. Together the complementary tasks provide adequate calories and diversified nutrition for the group as a whole. Of course, Neandertals were not tropical foragers, so the comparison is perhaps not apt. The Inuit are Arctic foragers whose diet of caribou, sea mammals, and fish is closer to the Neandertal diet than that of tropical foragers, yet Inuit have a division of labor that is, if anything, more distinct than that of tropical foragers. Instead of gathering edible plants, of which there are very few in the Arctic, women make and maintain much of the elaborate material culture that is essential to Inuit survival. This is a full-time undertaking that directly supports the hunting of the men.

Neandertals show a very different pattern of labor division; indeed, as we discussed in Chapter 2, it is difficult to find any evidence for a division of labor in the archaeological record.[18] For European Neandertals there is little evidence for the use of plants and small animals that would indicate economic activities complementary to large animal hunting. Nor is there any evidence for making or maintaining elaborate gear or clothing like that used by the Inuit. Neandertal equipment was pretty much the same everywhere, and they had no technologies that required long hours of manufacture and maintenance. As we have seen, there is some evidence for plant use by Neandertals, more in the Middle East than in Europe, but nothing to the extent shown by modern humans living on the same landscape after 20,000 years ago. The archaeological evidence really does suggest that all Neandertals were involved in the same economic activity: large mammal hunting. Within this activity they may well have had slightly different roles. Neandertal women show far fewer severe bodily injuries than the men, which suggests that they did not participate in the in-close killing of animals. But there are many other roles that they could have played, including driving animals or distracting them. And the women almost certainly participated in butchering and transport. Nevertheless, compared to modern hunter-gatherers, Neandertals had low levels of division of labor.

From the perspective of natural science, with its focus on reproductive strategies and mate choice, Neandertals' division of labor, or lack of it, has some interesting corollaries. As we have seen, evolutionary psychology has clearly established that men and women have different strategies of mate choice, which have selected for different criteria of attractiveness. Men tend to prefer young women at the peak of fecundity. Women are attracted to young, virile men, but they are also attracted to men who can provide for them and their children.[19] There are sound reasons for the difference, and it is clearly tied up in modern divisions of labor, at least for women. Could Neandertals have had a similar sex difference in criteria of attractiveness, given their very low levels of sexual division of labor? It's an intriguing question. We suspect that there must have been a difference in the sex difference. Neandertal men were almost certainly still attracted to young, fecund females; this is almost always a good male strategy. But Neandertal women probably had little reason to be attracted to older, established men. For one thing, there just were not very many of them. For another, those older men were probably *less* able to support a woman than a younger, healthier man. In other words, Neandertal men and women probably had similar criteria of attractiveness for members of the opposite sex.

From the social science perspective on sexual division of labor, with its focus on marriage, the implications are less clear. Given what we know about Neandertals, is there any reason to think that they even *had* marriage, a contractual agreement focused on children and wealth? This is a loaded question. Marriage of one kind or another is universal for modern people. If we deny it to Neandertals, they immediately appear less human. How can we even tell? How can we distinguish marriage from pair-bonding? One way is *through* division of labor, which is one of the core subjects of marriage contracts. This component of marriage agreements would appear to have been largely unnecessary for Neandertals; they all were involved in the same food-getting activities. We are left with the "children" component of marriage, in particular the potential for children and the affiliation of children. But this bears on who is related to whom in a social group, which is an issue we have not yet touched upon.

Finding a Mate

Evolution discourages inbreeding, or rather inbreeding discourages evolution. If closely related individuals mate, there are several negative consequences. First, the likelihood of their children inheriting a genetically determined birth defect goes up dramatically. It is not that the mating itself somehow does this, only that closely related individuals share a large number of genes, and if both carry a nasty recessive gene (that neither expresses), then one-quarter of their children, on average, will express the recessive trait. Thus over the long term closely related couples will be less successful reproducing than distantly related couples. Second, inbreeding reduces the variability of gene pools. Individuals become more genetically alike over time because people mate with people who are already genetically very like themselves. Natural selection operates on variability, and if a gene pool has little variability, it will be unable to adapt if circumstances change. This second problem may well have loomed large in the ultimate fate of Neandertals. These two biological problems mean that sexually reproducing organisms generally evolve mechanisms to reduce the possibility of inbreeding.

Monkeys and apes solve the inbreeding problem socially, often through outmigration. On reaching reproductive age either the males or the females leave their natal group or are driven out. They then must find another group to live with, or perish. Monkeys and apes cannot survive long alone; indeed the only anthropoid primate to spend extensive time alone is the male orangutan. Among Old World monkeys such as baboons it is the adolescent males who leave. They must attach themselves to a neighboring troop where they must fight their way up the dominance hierarchy if they hope to have any chance to reproduce. (If this sounds too much like high school, we apologize.) Because it is the males who leave, the remaining females are all related—mothers, daughters, sisters, and cousins of one another. Primatologists use the phrase "female philopatric" to identify such groups. Chimpanzees, on the other hand, are "male philopatric," which means that the males of a community are related, and the females leave at adolescence to join a neighboring community. Exactly how the outmigration occurs, whether by males driving off adolescents (baboons) or by females preferring foreign males (the Javier Bardem effect), varies a great deal among primate groups, but the biological effect is the same: avoidance of inbreeding.

There is a third problem created by inbreeding that complicates the lives of humans but is of no importance to monkeys or apes. Modern hunter-gatherers rely on social contacts with distant relations to buffer the consequences of local, temporary declines in the food supply. If the availability of important plants or animals declines, for whatever reason, hunter-gatherers can exploit the social obligations that their distant relatives have toward them, and of course vice versa. But if someone marries a sister or brother, they do not create this connection of social obligation. "Incest" is the term we use for the social prohibition against sexual relations with a close relative. It is a more subtle and complex concept than inbreeding precisely because it is not defined solely on the basis of biological nearness. In some societies, for example, one can marry cousins related through the father's sister, but not cousins related through the father's brother. Biologically they are equivalent, but socially they are not; the obligations one has toward a father's sister may very well differ from those one has with a father's brother. Modern humans, including hunter-gatherers, are very preoccupied with who can marry whom because it has such importance economically and socially. It is one way to create and maintain a robust social network. The universal incest taboo is as much about economic security as it is about successful reproduction.

What, if anything, can we conclude about Neandertal mating patterns and marriage? Because they were group-living primates, they must have practiced some kind of outmigration pattern in order to avoid the negative consequences of inbreeding. Either Neandertal boys or Neandertal girls left their natal community when they reached sexual maturity. But which? If Neandertals practiced a typical African ape pattern, and all hominins were ultimately African in origin, then they were male philopatric, with adolescent females moving to neighboring communities. In such a system, the men of a Neandertal group would be fathers and sons, brothers and cousins, and hence natural allies for cooperative action. This is perhaps the most probable scenario. Male philopatry is likely to be a deep-seated, phylogenetically constrained, social adaptation of the African apes and hominins that has selected for cognitive and behavioral dispositions. In less jargonistic terms, African ape males are prone to stay together, but females are attracted to foreign males. And male philopatry would have definite advantages in a situation of male provisioning. Because the males are related genetically, there is little evolutionary payoff to male-on-male aggression. But there is payoff to cooperation and sharing. If you are unsuccessful on a hunting foray, but your brother is successful and shares with you (and you with your mate), then your close genetic relatedness results in greater reproductive success for your common genes. Recently published evidence from the archaic DNA recovered from several Neandertal individuals from one site indicates that the males were related, but the females were not.[20] This is provocative evidence for male philopatry. There are, however, reasons for considering the alternative.

One of the unexpected facts provided by modern hunter-gatherers living in tropical Africa is the comparative unimportance of male hunting in providing calories for women and children. Men fail at hunting more often than not. When a hunt is successful, the meat is distributed throughout the group, so that everyone gets some. But it is not a reliable source of daily calories. These are supplied by the foods gathered by women. In a now famous study of the Hadza, who live in northern Tanzania, the anthropologist James O'Connell and colleagues demonstrated that men were not the most important provisioners of their spouses and children. The spouse's mother was.[10] Grandmothers, it turns out, are a more effective and reliable support for a young mother than is her husband. The mother's mother has a wealth of knowledge about the location of foods and the skill to acquire them. She is able to gather many more calories than she herself requires and, if she no longer has young children herself, can invest these calories in her grandchildren. But this works best if mothers and daughters stay together, in female philopatry. If a girl marries someone living in a foreign group, she will no longer live with her mother, but with her husband's mother, who will not be as invested in the well-being of her son's mate as in the well-being of her own son or daughter. (Like her son, she too will always lack 100% certainty about the paternity of her daughter-in-law's children.) If this reminds you of a divorce court plot, it is because provisioning and paternity are still topics of utmost interest in human lives.

Many readers will have noticed that we have avoided answering an important relevant question: Are modern people male or female philopatric? The answer is either or both. Anthropologists have been studying human kinship and marriage systems for a century and a half and have documented an almost bewildering variety of solutions to the problem of finding marriage partners. For some people (e.g., many Australian groups) the range of potential marriage partners is determined by elaborate rules that are sometimes so restrictive that there are few, and occasionally no, marriageable partners. For others, almost anyone of the opposite sex can qualify. (And recently an opposite-sex partner is not a necessity for marriage.) Anthropologists have tried in vain to identify causal connections between marriage systems and productive systems (hunting and gathering vs. farming vs. herding, etc.). There are some trends, but nothing strong enough to underpin an evolutionary argument. What we can conclude from this variety is this: humans are no longer naturally male or female philopatric, and this release from philopatry must have occurred at some point in our evolutionary past. But was it as long ago as *Homo heidelbergensis*, the common ancestor of Neandertals and us? If so, we would expect that Neandertals too came up with a variety of solutions. The status of Neandertals has some important implications for us; if they had a consistent system of philopatry, it would mean that the human release from philopatry evolved after the split.

We cannot, alas, say whether Neandertals were female or male philopatric, but the evidence, such as it is, suggests that they were one or the other. The archaeological evidence suggests that Neandertal communities were relatively small and thinly spread across the landscape. There is no evidence for regular interaction with neighboring communities, though there is evidence of occasional contact—the rare bits of high-quality raw material. Such low-intensity interaction argues against the presence of formal contractual marriage exchange systems, though de facto systems of outmigrating adolescent boys or girls are more likely.

Based on what we know of the archaeological record and the comparative evidence of nonhuman primates and human hunter-gatherers, we find the following sketch of Neandertal social groups to be most likely. Neandertals lived in communities of fewer than a hundred individuals, perhaps ten to twenty families related through the male line or the female line (male or female philopatric). Based on movement of raw material, Neandertal communities in western Europe occupied a territory of perhaps 2,000 square kilometers; larger territories characterized Neandertals in eastern Europe, perhaps because of a different distribution of game. The families consisted of a pair-bonded man and woman, their children, and perhaps a surviving older relative. These families were the primary foraging unit, with all members participating in the hunt. Each family may often have foraged with another family. The several families of a community came together when big kills were made (e.g., at La Cotte) and stayed together as long as the meat lasted. Members of the community had fairly regular contact with one another, but little contact with neighboring communities. The Neandertal community was exogamous; either boys or girls left their natal community to join a neighboring community. Exactly how this played out is beyond our ability to detect. The only possible traces of such outmigrations are the small quantities of high-quality raw material from sources well beyond the usual community range, carried in perhaps by an adolescent boy or girl to defuse the natural aggression elicited by strangers.

Social Roles

Beyond age and gender there is little evidence for Neandertal social roles. We have no evidence that adult women or men ever differed from one another in status or in range of activities. It is likely that individuals differed in dominance, but such differences were not marked in a way visible to archaeologists, if they were marked at all. In such small communities everyone is intimately acquainted with everyone else, and it is unnecessary to mark one's rank.

Social Cognition

Neandertals lived their lives in social groups. The bias in the paleoanthropological record toward tools and the quest for food must not blind us to the importance of social interaction in the lives of Neandertals. Indeed social interactions of all kinds were Neandertals' paramount concern, as they are for modern humans and nonhuman primates. And much of Neandertal thinking must have been geared to the successful performance of their social lives. But how can we access the part of their mind devoted to such thinking? Archaeological clues to the social lives of Neandertals are less direct and less abundant than the evidence for technology and subsistence, and our reconstruction of Neandertal social cognition must therefore rely more on our understanding of the social cognition of modern people and, to a lesser extent, nonhuman primates. We will try to identify components of Neandertal social cognition by asking what Neandertals needed to know and understand in order to perform the social interactions apparent from the archaeological record. As always, we will be conservative, identifying the minimum cognitive abilities required for a particular interaction. And we will be conservative in an additional sense: our interpretations will be consistent with what we already know about Neandertal cognition from technology and subsistence. We will first address cognitive demands of Neandertal face-to-face groups, after which we will look at additional requirements presented by the territorial community.

The most likely makeup of a Neandertal foraging group was an extended family, and at the core of this family was a pair-bonded Neandertal woman and man. One of them, probably the woman, had grown up in a neighboring territorial community and migrated in at adolescence. How did a man and woman, strangers to one another, form the social bond of trust necessary to act as a team? Initially the bond was likely to have been sexual and romantic. Sexual and romantic attraction is a powerful drive in modern human lives, and it has a strong cognitive component for which we have many evocative metaphors: love is blind, a many splendored thing, and so on. Initially, as for us, sexual attraction was the glue that held the Neandertal adult pair-bond together. Emotional bonds undoubtedly linked all of the members of a Neandertal extended family, including children and any attached adults, who had family bonds to either the man or the woman. The archaeological evidence points to a second component of Neandertal pair-bonding: empathy. Neandertals cared for injured members of their family groups, and also protected corpses. This suggests an ability to feel the pain of another person, to imagine the needs of another, and try to alleviate them. This requires Theory of Mind and is entirely consistent with the levels of intentionality we identified in Neandertal technical cognition. Finally, because Neandertals were primates, and because they were similar to us, we must add a dash of dominance to the family social mix. Some of the social interactions within Neandertal families were based on power, the threat of violence, and occasional actual violence. Cognitively this invokes the manipulation of emotional responses such as fear and remorse. Neandertals undoubtedly lost their tempers and struck out at family members, but they also undoubtedly had ways to defuse aggression and establish harmony. From this brief description Neandertals appear to be very much like us, and indeed they were. The cognitive basis of such highly charged face-to-face interactions, for both Neandertals and for us, is actually linked intimately to our use of our own bodies and depends far less on symbols and language than you might suppose.

Much of small-scale social cognition is embodied.[21] We introduced the notion of embodied cognition in the previous chapter on Neandertal technology, where we argued that bodies and tools are extensions of the mind itself, not just for Neandertals, but for ourselves. Bodies similarly extend the mind in social interaction; much of our hourly, daily social interaction is done with our faces, hands, postures, and so on. We have all had the experience of walking into an occupied room and knowing immediately that something was amiss. How did we know? We knew because much of our knowledge of social situations comes from the postures and facial expressions of others and ourselves. As primates, we have evolved neural mechanisms to support this ability. One such ability is motor mimicry. When interacting with others, we tend to take on the same postures, facial expressions, and action patterns. We cannot help doing this. What happens when someone in a group yawns? Almost always, others in the group will feel the urge to yawn. This is a trivial, well-known effect, but often such postural or motor entrainment is not trivial. If you assume an aggressive, threatening stance, and the person you interact with does the same, the result is an escalation of hostility. (And if you have been drinking, which lowers inhibition, the result may be violence.) Alternatively, if you take an open, vulnerable posture, you may well defuse a potentially hostile exchange. But embodied social interaction can be more subtle. You can subtly change a social dynamic by changing your facial expression. We can all feign surprise or boredom, or even love if the situation warrants it. And we need not say a word. In other words, we respond bodily to social situations and can also manipulate social interactions bodily.

The ability to interact socially via facial expressions, postures, and positioning is for the most part not learned. We do not need to be taught what anger looks like, or affection, or indeed the entire gamut of emotion. We do need to be taught when and how to suppress or exaggerate expressions and postures, and, interestingly, such social conventions vary a great deal among modern people. But the basic social postures do not. They are the product of millions of years of primate and hominin evolution, and they are very effective tools of social interaction. Undoubtedly Neandertals had them and used them. Neandertal pair-bonding played out via the same postures and expressions we also use, including touching, smiling, embracing, and entraining postures and expressions. Neandertals acted out dominance in aggressive stances, closing personal distance, and violent displays (throwing objects, etc.), just as we do, just as nonhuman primates do. These social abilities are most effective in precisely the social circumstances in which Neandertals spent the majority of their lives: small face-to-face groups of closely related people. The interesting question is not, Did Neandertals rely on embodied social cognition? They must have. The interesting question is, Did they rely on anything more? To answer this we need to consider Neandertal territorial communities.

We know that Neandertal foraging groups occasionally congregated into larger assemblies, most likely encompassing other members of a territorial community related to one another through either the men or the women. The archaeological evidence does not reveal how often this happened, but it does suggest something about the circumstances. These larger groups congregated at kill and butchery sites such as La Cotte and Salzgitter-Lebenstedt. There is little good evidence of large congregations in any other circumstances. The largest possible habitation site is the one at Molodova, where, if all of the purported huts or windbreaks were occupied at the same time, there may have been twenty to forty people, almost certainly fewer than the total community. Most of the time Neandertal foraging groups appear to have lived separately. In other words, Neandertals rarely, if ever, came together except in the uncommon cases of large-scale butchery events. There is no evidence that Neandertal territorial communities ever came together for other reasons. There were no social events that required the presence of all members of a territorial community—no weddings, or funerals, or initiations.

Large social aggregations present cognitive challenges, even (or perhaps more so) if everyone is related, or known, to everyone else. This is one reason weddings present so much fodder for comedy and tragedy in the modern world. The more people there are, the more relationships there are to be managed, especially if the people "have history." And everyone in a Neandertal territorial community would have had history. When modern hunter-gatherers are asked why groups fission, they cite interpersonal conflict as the most important reason. Yet modern hunter-gatherers are also pulled together by requirements of community action in the form of weddings, initiations, and so on. This centripetal force was not a feature of Neandertal society. The larger the interacting group, the harder it is for the resources of embodied cognition—posture, gaze, touch, and so on—to maintain social bonds.

According to the primatologist Robin Dunbar, this was a major reason for the evolution of hominin cognition and language.[22,23] Large group interactions are partly a challenge for long-term memory and partly a challenge for working memory. We suspect that Neandertals' long-term memory was easily capable of remembering everyone in their territorial community and the specific relationships held with each one. It is monitoring and maintaining all of them in a group setting that requires working memory, more specifically Theory of Mind. (Think of being at a party trying to remember what you have said to whom, and whether the man or woman you are talking to has already spoken to someone else you have spoken to.) Neandertals' apparent reluctance to congregate long in large groups may reflect an aversion to placing such a load on working memory. The reader may well counter at this point that many nonhuman primates, such as baboons, congregate in permanent cohesive troops larger even than Neandertal territorial communities, and do not have expanded working memory capacities. This is true, but these monkeys congregate for sound ecological reasons—predator detection and avoidance—and resort to very strict dominance hierarchies to organize most of their interactions. Neandertal ecology did not favor large cohesive groups, and in the absence of such a centripetal force, the cognitive challenge of large social congregations may have been enough to limit them to task-specific events such as butchery. In other words, we suspect that Neandertals relied on their resources of embodied social cognition to maintain and establish social relationships, and that this was effective for most of their daily lives in the context of small groups. They also relied heavily on these resources when congregating in larger task-focused assemblies where specific local goals temporarily muted the opportunity for social interaction. But Neandertals were never really called upon to use more elaborate forms of social cognition.[24]

What's Missing?

The picture of social cognition presented to us by Neandertal archaeology is one of small-scale, intimate interactions and the cognitive abilities those required. Modern people live in larger face-to-face groups and also interact regularly with people in other face-to-face groups. We negotiate our lives not just with extended families, but also with acquaintances and even strangers. And we appear to have cognitive abilities that evolved to cope with such interactions. Evolutionary psychology and cognitive social psychology are just beginning to identify and confirm these abilities, but already two stand out as important components of the modern mind.

Cost-Benefit Analysis

Over the past twenty years the anthropologist Alan Page Fiske has developed a sophisticated model of social cognition based on "relational models."[25,26] The four relational models were identified by a well-established tool known as factor analysis, which identifies traits that covary (i.e., change together), suggesting a common source, in this case a common cognitive basis. Fiske argues, and much research backs him up, that these four relational models are discrete (separate) cognitive ways of organizing social behavior, and that they account for virtually all human social interactions. He also argues that they are universal in modern people, and that young children "innately expect" them to be in play in any social interaction. The four are *communal sharing*, which operates largely through embodied processes; *authority ranking*, which mediates dominance relations; *equality matching*, a process to attain balance in relationships; and *market pricing*. The first three all have roots in nonhuman primate social cognition and were all undoubtedly components of Neandertal social cognition. The fourth one may be unique to modern humans.

Market pricing is the ability to make proportional transactions in which commodities of one kind are traded for commodities of another kind. Most often, people in today's world do this through the abstract, symbolic medium of money, which may have no inherent value but has significant symbolic value. But currency is not the only way people negotiate commodities. Any time we exchange one kind of thing for another kind of thing—money for food, labor for food, beads for land, etc.—we must calculate a ratio of value. We must make a cost-benefit analysis. This kind of thinking is the basis of contracts and all economic exchange. It is most efficiently done with numbers and systems of calculation, but these are not necessary. The ability to think proportionately is. And according to Fiske, we have evolved cognitively to be quite good at it. Such cost-benefit thinking would not have been required for Neandertals' small-scale, intimate groups. It is the kind of thinking we use when negotiating transactions with acquaintances and strangers, and, as far as we know, Neandertals never did this. Dealing with acquaintances and strangers also dramatically increases the opportunities for cheating, and we have also evolved an ability to cope with this.

Cheater Detection

Imagine for a moment that you are a police officer trying to enforce the rule that an individual must be 21 years of age to drink alcohol. You are called to a party where some people are drinking beer and some are not. How do you assess whether the rule is being violated? It should be obvious that you would check the age of people who are drinking, and if some are under 21, they have violated the rule: they have cheated. The logical structure of this bit of law enforcement is a classic conditional of the "if P, then Q" variety: if you are drinking, then you must be 21. Notice that there are four kinds of people at the party: those who are drinking and are 21, those who are not drinking and are not yet 21, those who are 21 and not drinking, and those who are not 21 but are drinking. Most people presented with this problem have absolutely no difficulty figuring out that it is the last group that must be identified in order to demonstrate that the party is in violation. However, if you present the same situation as purely a logical problem, the result is quite different. Instead of a party and drinking, let's present our problem in the classic "if P, then Q" logical form. Each of the following cards has a condition of P and a condition of Q (taken from P, not P, Q, and not Q), one visible, one written on the back:

If given two choices, which cards would you choose to test the rule (if P, then Q)? It's not so easy now. Indeed, only 5 to 25 percent of people tested pick "P" and "not Q," the correct response. Why should that be? The logical structure is the same. It turns out, as the evolutionary psychologist Leda Cosmides has clearly established, that the problem is content-dependent.[27] When presented in the form of a social contract—in order to have P, you must be Q—people generally have no difficulty solving the problem. But in other content situations, as in formal logic, they do not do as well.

Do these experimental results represent just a quirk of human reasoning, or do they represent something more profound? Cosmides argues that evolution has selected for this content-specific form of thinking, but not a more general "conditional" problem-solving ability. Human life is riddled with agreements made with other people, and anyone who is able to detect cheaters will be able to avoid costly choices (real and metaphorical) by either avoiding the cheater in the future or punishing the cheater now. There is no similar advantage to general conditional reasoning, so it must have arisen in the narrow domain of social cognition. But when did it evolve? We suspect that Neandertals did not possess this ability. The ability to detect cheaters is most useful in contractual interactions with acquaintances and strangers. In the territorial communities in which Neandertals spent their lives, everyone knew everyone else intimately. And even if some individual Neandertal were prone to laziness or not pulling his or her weight, this character flaw would soon be known to everyone. Moreover the kinds of economic dependencies true for Neandertals were just not the kinds of contractual rule-based obligations so common in the modern world. Neandertals relied on families and the emotional bonds that bound them together. There was no need for cheater detection, which became advantageous only when nonemotional and contract-like agreements between acquaintances and even strangers became common. But if Neandertals lacked this ability, what might have happened when they did encounter strangers who were adept at social and economic contracts? We suspect that Neandertals' direct, emotional, embodied style of social cognition placed them at a marked disadvantage. In fact this may have played a role in their ultimate demise. The modern humans who entered central and western Europe beginning 40,000 years ago lived in larger face-to-face groups than Neandertals did, maintained regular social contacts with acquaintances who lived hundreds of kilometers away, and almost certainly had the ability to negotiate with strangers. Neandertals would not have known how to respond.

The culture of these modern humans included the extensive use of symbols, and the contrast between Neandertal symbol use and modern human symbol use may also have played a role in the fate of Neandertals.

Further Reading

1. Gamble, C. 1999. *The Palaeolithic Societies of Europe*. Cambridge, U.K.: Cambridge University Press.
2. Mellars, P. 1996. *The Neanderthal Legacy: An Archaeological Perspective from Western Europe*. Princeton, N.J.: Princeton University Press.
3. Stringer, C. B., and C. Gamble. 1993. *In Search of the Neanderthals: Solving the Puzzle of Human Origins*. New York: Thames and Hudson.
4. Trinkaus, E., and P. Shipman. 1992. *The Neandertals: Changing the Image of Mankind*. New York: Knopf.
5. Jolly, A. 1985. *The Evolution of Primate Behavior*. New York: Macmillan.
6. Ingold, T., D. Riches, and J. Woodburn, eds. 1988. *Hunters and Gatherers*. New York: Berg.
7. De Lumley, H., ed. 1969. *Une Cabane Acheuleenne dans la Grotte du Lazaret*. Paris: Memoires de la Societe Prehistorique Francaise 7.
8. Hoffecker, J. 2002. *Desolate Landscapes: Ice-Age Settlement in Eastern Europe*. New Brunswick, N.J.: Rutgers University Press.
9. Goodall, J. 1986. *The Chimpanzees of Gombe*. Cambridge, Mass.: Belknap Press.
10. O'Connell, J., K. Hawkes, and N. Burton Jones. 1999. Grandmothering and the evolution of *Homo erectus*. *Journal of Human Evolution* 36:461–485.

11. Hawkes, K., and R. Paine, eds. 2006. *The Evolution of Human Life History*. Sante Fe, N.M.: School of American Research.

12. Guatelli-Steinberg, D., D. Reid, and T. Bishop. 2007. Did the enamel of Neandertal anterior teeth grow differently from that of modern humans? *Journal of Human Evolution* 52(1): 72–84.

13. Smith, T. M., et al. 2010. Dental evidence for ontogenetic differences between modern humans and Neanderthals. *Proceedings of the National Academy of Sciences* 107(49): 20923–20928.

14. Rosenberg, K. 2004. Living longer: Information revolution, population expansion, and modern human origins. *Proceedings of the National Academy of Sciences* 101(30): 10847–10848.

15. Trinkaus, E. 2011. Late Pleistocene adult mortality patterns and modern human establishment. *Proceedings of the National Academy of Sciences* 108(4): 1267–1271.

16. Fox, R. 1983. *Kinship and Marriage: An Anthropological Perspective*. New York: Cambridge University Press.

17. Anton, S. 2003. Natural history of *Homo erectus*. *Yearbook of Physical Anthropology* 46:126–170.

18. Kuhn, S., and M. Stiner. 2006. What's a mother to do? The division of labor among Neandertals and modern humans in Eurasia. *Current Anthropology* 47(6): 953–980.

19. Buss, D. 2003. *The Evolution of Desire: Strategies of Human Mating*. New York: Basic Books.

20. Lalueza-Fox, C., et al. 2011. Genetic evidence for patrilocal mating behavior among Neandertal groups. *Proceedings of the National Academy of Sciences* 108(1): 250–253.

21. Robbins, P., and M. Aydede, eds. 2009. *The Cambridge Handbook of Situated Cognition*. Cambridge, U.K.: Cambridge University Press.

22. Dunbar, R. 1993. Coevolution of neocortical size, group and language in humans. *Behavioral and Brain Sciences* 16:681–735.

23. Dunbar, R. 2009. Why only humans have language. In *The Prehistory of Language*, eds. R. Botha and C. Knight. Oxford: Oxford University Press.

24. Ambrose, S. 2010. Coevolution of composite-tool technology, constructive memory, and language: Implications for the evolution of modern human behavior. In *Working Memory: Beyond Language and Symbolism*, eds. T. Wynn and F. L. Coolidge. Chicago: University of Chicago Press.

25. Fiske, A. P. 2004. Four modes of constituting relationships: Consubstantial assimilation; space, magnitude, time, and force; concrete procedures; abstract symbolism. *Relational Models Theory: A Contemporary Overview*, ed. N. Haslam. Mahwah, N.J.: Lawrence Erlbaum.

26. Haslam, N., ed. 2004. *Relational Models Theory: A Contemporary Overview*. Mahwah, N.J.: Lawrence Erlbaum.

27. Cosmides, L. 1989. The logic of social exchange: Has natural selection shaped how humans reason? Studies with the Wason selection task. *Cognition* 31:187–276.

5

It's Symbolic

ANTHROPOLOGISTS LIKE TO say that modern life is immersed in symbols. What do they mean? For semioticians, the academics who study such things, a symbol is a variety of sign, and a sign is something that can "stand for" something else.[1] Imagine that you are driving down a highway and see a square piece of metal, painted yellow, mounted on a post with an image of a stag. What does it mean? You know that it is a warning to watch out for deer; the piece of metal has come to "stand for" deer. But this example is not a true symbol in the semiotic sense because the "reference" of the sign is delivered by a facsimile, a likeness, of the referent (deer). It is an icon. To be a true symbol the reference must be arbitrary, which means that there is no natural link between the sign and what it refers to. The word "deer" does not resemble a deer; it is a true symbol. Somehow a word like "deer" comes to stand for a object like deer. There is a third variety of sign, the index, which produces reference through association; smoke is an index (stands for) fire, and a hoof print can be an index for deer. Anthropologists often use these semiotic definitions, but they also occasionally lump them all together and refer to the bunch as "symbolism." When anthropologists say that modern life is immersed in symbols, they usually mean the whole package.

But what does "immersed" mean (other than that anthropologists are all wet)? In the modern world we have symbols all around us.[2] Most obvious are words (more on words in the next chapter), but there are many other kinds. We once attended a reception at the home of our university chancellor. In making some welcoming remarks, he observed that he knew it was a reception for faculty because "all of the cars outside are ten-year-old Toyotas." Most of you get the joke, we're sure, but what does symbolism have to do with it? It is symbolism because the choices people make can come to "stand for" them; in this case purchasing Toyotas and keeping them for years is the kind of choice college professors tend to make. In one sense these Toyotas are indexes; the economic status of college professors places Toyotas in their buying power, and Toyotas' reliability makes them a good long-term investment. There is a natural association. But there is also something a bit symbolic too. When choosing a car to buy, people also send a message to the community, often quite consciously, about who they think they are. Few college professors purchase Ford F250 pickup trucks with lift kits and monster tires, even if they live down a perpetually muddy lane. It just sends the wrong message. It is this message role of symbols (in the general sense) that is the heart of the "immersion" emphasized by anthropologists. We are all very much attuned to the messages we and others send through the clothes we wear, the places we live, the tools we use, the jobs we perform, and so on. Many of these messages are conscious—most speech, advertising, stories—but much is largely unconscious.

Not only are symbols all around us, but they have come to be one of the primary vehicles for learning. Again, most obvious are the words of language and, in the contemporary world, written words and mathematical symbols. More subtle and profound information is delivered through narratives—stories and parables—and visual symbols that are loaded with meaning. Think of a Christian crucifix, and the information and emotion packed into it. It is an icon (image of Jesus), an index (associated with the passion of Christ), and a symbol (the resurrection, hope for eternal life, etc.). The last meaning represents yet another way that anthropologists often use the term "symbol." Such symbols are the only way one can think and talk about abstract notions and ideas. The references to a person (Jesus) and an event (the crucifixion) are fairly straightforward signs standing for an individual not present and an event in the past. But "resurrection" is not a simple reference. What is a resurrection, anyway? It is certainly not a feature of the tangible, visible world. It is built out of abstract, intangible ideas, and you can only think about or talk about such things using symbols. When anthropologists (and archaeologists) refer to symbolic culture, they almost always mean a knowledge system that incorporates such abstract notions. All modern people have symbols that are equal in subtlety to the Christian crucifix. But what about Neandertals? Some paleoanthropologists suspect that Neandertals may not have been "immersed" in symbols to the same degree.

Consider for a moment the Hohlenstein-Stadel figurine (Figure 5-1).[3] What do you see? Most people see a representation of an imaginary creature with a human body and a lion's head. Such a creature is a product of pure imagination; it does not exist in the real world, and never has. The statuette is

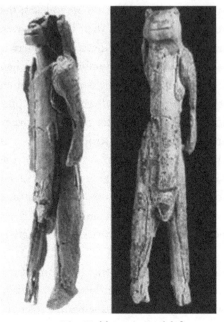

FIGURE 5-1 The Hohlenstein-Stadel figurine, an object that is almost certainly an artifact that had an abstract, symbolic meaning (after Wynn et al.³). It was made by modern humans about 30,000 years ago.

a depiction (icon) of an abstraction. Interestingly most people to whom we have shown pictures of this figurine immediately conclude that it must have had religious significance, primarily because in our world such creatures often do. But not always. Think of a werewolf; it is not really a religious concept (not Christian anyway), but it certainly is an abstract notion with a very real ability to frighten, and perhaps instruct ("Don't play in the woods at night when the moon is full"). So it is symbolic, as was Hohlenstein-Stadel for its makers. We do not know *what* it stood for, but we know *that* it stood for something. This figurine is about 32,000 years old and was made in Germany by modern humans. Its presence tells us that its makers used symbols in ways very similar, if not identical, to the ways we use symbols.

The archaeological record of Neandertals provides us with nothing comparable to Hohlenstein-Stadel, but to be fair we must also note that such unambiguous examples of symbols are pretty rare for modern humans until well after 30,000 years ago. Moreover if Neandertals made symbols out of perishable material (e.g., wood) we might never find them. As the adage goes, absence of evidence is not evidence of absence. On the other hand we cannot attribute symbolic culture to Neandertals just because it would be fair. There needs to be evidence, even if it is indirect. The archaeological evidence of Neandertals does include three kinds of evidence that have possible implications for symbolic culture: burial, use of fire, and body ornamentation.

Mortuary Treatment

What do you do with dead people? If people are dead, are they still people? Certainly the recently dead (one can't help thinking of Marty Feldman in the brain lab in *Young Frankenstein*) elicit emotional responses from the living that are as intense as or even more intense than the ones elicited in life. Their altered status leaves a social gap that is, for a time at least, very real. Indeed you could argue that at least a few of an individual's social roles continue after death, especially the links in the social web that connect other people through the point that was that person. Death is a serious issue for modern people, but was it as serious for Neandertals?

Mortuary practice is another of those topics, like marriage, that has fascinated anthropologists for over a century and a half. It has also fascinated archaeologists, who occasionally encounter mortuary treatments in their excavations, and psychologists, who have delved into the role of death in the makeup of the modern psyche. There is, as you might guess, an immense literature on death, which we will not attempt to summarize. But from the anthropological perspective at least, there are several features of death and mortuary treatment that all modern people share and that can serve as a framework to assess Neandertal practices.

First, *emotional saliency*: death is emotionally traumatic. There is the fear, some would say terror, that we all have for our own death. We may "repress" this fear, to use an old term from psychoanalysis, especially in our daily lives, but it is always lurking in the back or our mind. This fear is very old in an evolutionary sense, and powers important preconscious responses, such as fight or flight. It is true that many people actively acknowledge and discuss their own death and place it in a well-developed symbolic or religious context, but the fear of death is still there. The loss of a loved one elicits grief, an emotion so strong that it causes physiological responses that extend beyond the brain itself. And the closer the emotional bond, the stronger the grief. Because modern people often establish pair-bonds with their mates, grief extends to the loss of a wife or husband.

Second, there is *social disruption*: death leaves a hole in the social fabric. The death of an adult especially requires a realigning of social and economic relationships. The tasks performed by a spouse in the division of labor must now be performed by someone else. There is also the problem of what to do with surviving family members, both near and distant. And in some cases death can have significant political consequences.

Third, we see *emotional ambivalence*: the death of an adult is often both a blow and a boon to survivors.[4] Children or siblings may inherit wealth or power from the deceased, or may be freed from emotional and familial obligations that have been a burden for years. This combination of grief and joy makes the death of an adult a source of complex feelings and responses.

Fourth, there is *continuance*: with very few exceptions, modern people believe in the continuance of a person beyond death. This takes many different forms, from the idea that souls are separate from bodies and continue in a parallel but separate existence, to notions of ghosts, zombies, and other spirit entities that exist in our world, or even reincarnation.

Fifth, *ritual mitigation*: without exception people cope with the emotional and social trauma of death through ritual performances. In anthropological parlance such rituals are "rites of passage," performances that mark the major transitions in an individual's life and role in a community.[5] Rites of passage follow rules that are often very strict and guide the participants through the emotional trial of transition. Death rites can be elaborate, time-consuming, and expensive, or simple, brief, and cheap, but they are always put into practice. Even in our cynical modern age we follow ritual treatments of death in the form of memorial services, moments of silence, donations to charities, and spreading ashes, because the emotional and social disruptions of death have not gone away. Death rituals often perform two seemingly contradictory functions: severance and integration. The dead person must be severed from the living; everyone (including the deceased) must know that he or she is truly gone and can no longer act in the world as in the past. At the same time this dramatic change in status requires that the deceased be either reintegrated with the living as a changed, ancestral presence, or integrated with some other existence beyond death. Mortuary rituals often require extended periods of time (days, weeks, sometimes even longer) during which survivors have a temporary changed status—mourning—that sets out rules for action that mitigate the emotional impact of the death.

Sixth, *transformation of the corpse*: no one simply carries a corpse out back and leaves it with the rubbish. Treatment of the corpse is an important part of mortuary ritual and is often tied to ideas of continuance, severance, and integration.[6] For many people, complete severance of the corpse from the living is essential, hence burial away from living areas or cremation (transforming the body to ash). But for many people it is important that a transformed corpse be reintegrated with the community. In Neolithic Orkney just 5,000 years ago, people practiced excarnation—exposing corpses until carrion-eating birds had stripped the bones—after which people moved the bones into communal stone tombs that became places of recurrent rituals.[7] These Orcadians clearly wanted to maintain a connection. In the American Southwest, on the other hand, Navajo carried corpses away from settlements and houses and disposed of them in hard-to-reach places. Here severance was essential. Sometimes people erect elaborate visual markers memorializing the individual (occasionally lacking a corpse entirely), and sometimes the dead receive no marking at all; they are "forgotten" after severance from the living and never mentioned again. For many people in the modern world, death is not final until the final transformation of the corpse has been achieved.

Seventh, there is *variability*: modern people have devised a large variety of mortuary rituals and accompanying corpse treatments. These are so varied that it is difficult even to connect specific corpse treatments to specific beliefs about the dead. Some cremation, for example, marks complete severance of the corpse, and the dead from the living, but sometimes cremated remains are lovingly collected and rehoused with the living. The only common denominators are the universality of ritual of some kind and corpse treatment of some kind.

Grave Miscalculations

One of the few commonly known facts about Neandertals is that they buried their dead. Some of the first Neandertals ever found over a century ago in southwestern France were purported to have been found in graves, including the Old Man of La Chapelle. For the original French archaeologists burial carried with it an implication of ritual corpse treatment, and ritual in turn implied religious sensibility, perhaps including a concept of afterlife. In other words, they concluded that Neandertals shared with us a sense of the supernatural. Like many established truisms, these ritual graves of Neandertals have wielded an unfortunate influence on our understanding of Neandertal lives, implying an emotional and religious similarity to us that is overstated at best, and perhaps even entirely unwarranted. Both terms, "ritual" and "grave," are misleading. We have no evidence that ceremony of any kind accompanied corpse treatment, and the small-scale scrapings of earth qualify as graves in only a very technical sense. So just what were Neandertals doing with corpses?

Perhaps the most famous Neandertal mortuary site of all is the cave of La Ferrassie in southwestern France. Between 1909 and 1921 archaeologists recovered the remains of seven Neandertals; two of them consisted of nearly complete skeletons, and the other five were partial skeletons. Introductory textbooks on human evolution still commonly reference these La Ferrassie Neandertals and the graves that contain them, and often include a photo of La Ferrassie #1 (a Neandertal with a huge cranial capacity of 1,700 cubic centimeters). But close examination of the sources reveals something odd: La Ferrassie #1 was *not* found in a grave. It lay in a depression next to the back wall of the cave, its head and shoulders covered by three limestone blocks. La Ferrassie #7 was similarly situated nearby. The legs and arms of the skeletons were slightly bent ("semiflexed" is the archaeological term), but not enough to suggest that the corpses had been bound into tight positions ("flexed"). Indeed the only possible graves at La Ferrassie were two shallow pits in which #3 and #4 were found. The pits were very small, from about 1 to 2 ¼ feet deep, and the skeletons in them only partial (#3 had only arms and a head; #4 was actually two fragmentary skeletons of a 10-month old and a neonate). In other words, not one of the La Ferrassie Neandertals was an intact burial. The most complete, #1 and #2, were not buried, and those in the only possible graves, #3 and #4, were incomplete. This is hardly convincing evidence of mortuary ceremonialism. But La Ferrassie was excavated long ago, when excavation standards were a bit more lax (but not by much). Perhaps a more recent excavation will be more definitive.

Kebara Cave in Israel, mentioned earlier in the context of Neandertal hunting and plant consumption, provides the best recently excavated example of a Neandertal grave.[8] All told, Kebara has yielded the remains of at least twenty-three adult and juvenile Neandertals, most represented by scattered teeth. But two of the Neandertals appear to have been buried. The better of the two examples is Kebara #2 (#1 was an infant excavated in the 1960s). Kebara #2 lay on its back in a shallow, basin-shaped pit that truncated earlier deposits. This truncation is important because it indicates that the pit had cut into sediment that had been laid down earlier, one the hallmarks of a grave (Figure 5-2).

When one of us (TW) first saw this section drawing years ago, he said aloud, "Looks like a grave to me." But closer consideration revealed some oddities. First, the grave is very shallow, only 20 to 25 centimeters deep (about 10 inches). Second, the

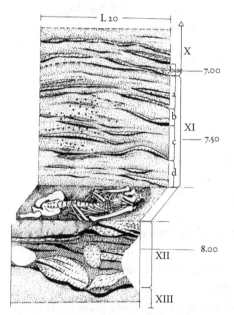

FIGURE 5-2 Profile drawing from Kebara Cave showing the stratigraphic position of the Kebara #2 skeleton, lying in a shallow depression (after Bar Yosef et al.[8]). The black and hatched bands mark the positions of hearths.

cranium was missing, and indeed could never have been in the pit, which was too shallow. But the jaw was there. Either the corpse was buried with the head exposed, and the cranium removed later, or it was never buried at all. Third, the right leg is entirely missing, as is most of the left. And fourth, archaeologists could find no evidence for one side of the pit wall, suggesting that the "pit" may simply have been an erosional depression. This would perhaps not be too troubling, if it were not that Kebara #2 is about the best example of a Neandertal grave that paleoanthropology has to offer.

Twenty years ago the archaeologist Rob Gargett took a critical look at the evidence for Neandertal burial and found it seriously deficient.[9] He found instead that the graves in sites such as La Ferrassie, La Chapelle-aux-Saints, Shanidar, and several others could all be explained as natural processes, not intentional burial. Professional reaction to this careful bit of scholarship was swift and harsh. You would have thought Gargett had questioned the validity of the double-helix model of DNA. Clearly, his detractors argued, his standards were too high and his understanding of French too clumsy, and he had avoided discussing the best examples. Moreover you could hardly expect the earliest evidence for burial to be unequivocal. But Gargett stuck to his guns, and in 1999 responded to his critics with a second article critiquing sites he had not discussed in the initial publication.[10] His conclusion was the same: there are no convincing examples of Neandertal burials.

Only circumstantial evidence prevents us (and others) from embracing Gargett's conclusion in its entirety; there are just so many relatively complete and partial articulated Neandertal skeletons. "Articulated" means that the bones of the skeleton lay in, or very near, their natural anatomical position in relation to one another. In other words, the bodies had been intact prior to being covered by sediment. Gargett does a good job of explaining, in each case, how natural processes of sedimentation might have entombed the corpses, but the fact remains that the sheer number of articulated skeletons and partial skeletons contrasts starkly with the record of earlier hominins, and even contemporary *Homo sapiens* living in other parts of the world. Something was different about Neandertals. It could be as simple, as Gargett suggests, as that Neandertals lived in caves more often than others did, and hence died in caves more often, or that there were no carrion-eating carnivores (e.g., hyenas) living in the region in which we find the burials.

We think that a third interpretation of the evidence explains it better: Why not accept that the ambiguity of Neandertal mortuary practice is an accurate reflection of Neandertals' investment in corpse treatment? In this scenario Neandertals' mortuary practices were real but minimal—minimal to the degree that only fifteen to twenty skeletons have managed to survive the ravages of time reasonably intact. The archaeological evidence itself supports this. There are few, perhaps no, real graves in the sense that we think of them. Most often Neandertals simply took advantage of a natural depression, such as the one at the rear of La Ferrassie, or perhaps expanded a natural low spot by scraping out some earth. They then positioned the corpse, bending the limbs a bit if necessary, and perhaps covered it with some earth or debris. That's all. It is easy to see that the vast majority of such treatments would not survive long. We suspect, for example, that all twenty-three Neandertals at Kebara received this kind of treatment, and that twenty-one of them were quickly disturbed and taken apart by animals and geomorphological processes. By sheer good fortune Kebara #2 survived, without its head and lower limbs. Neandertals did not just leave corpses where they died. They moved them and invested labor in protecting them, but there is no reason to conclude that they did anything more than this.

At Moula-Guercy, a cave site along the Rhone River in southern France, Neandertals practiced a very different form of corpse treatment: cannibalism.[11] The site was occupied about 100,000 years ago during a relatively warm climatic phase. In size and layout it resembles most other Neandertal cave sites, a place where a small group resided off and on, making and tending fires and butchering and cooking animals they had killed. But among the butchered carcasses of red deer and goat were the butchered carcasses of six Neandertals. The bones have the same kinds of butchery marks as the other animals, including cut marks where a shoulder had been cut from the torso, others where the Achilles tendon had been sliced through, and others where an elbow had been cut apart, a tongue had been cut out, and the temporalis muscles stripped from the skull. The thigh bones had been smashed to get at the marrow, and the skulls shattered to access the brains. It was an efficient job of butchery, with cooking and eating the ultimate goal. Of the six Neandertals, two were adults, two were teenagers, and two were children between 6 and 7 years old, possibly the remains of all or most of a single Neandertal residential group.

It is unlikely that such a group died natural deaths, and by extension unlikely that this was a form of ritual corpse disposal. There are several well-documented examples of ritual cannibalism in the anthropological literature on modern people, and modern participants provide a variety reasons for it, such as a final insult to an enemy or a way to obtain power, or as a way of continuing the essence of an individual in his or her surviving relatives. While it is possible that one Neandertal group killed and ate another in order to acquire the power or spirit of enemies, a practical explanation is more likely: one group killed and ate the other because they were hungry. The victims may have been strangers, not members of the butchers' territorial community, and therefore fair game. Perhaps a hunting party of foreign Neandertals ambushed the Moula-Guercy group in their residential cave. This would explain the age distribution of the victims. Archaeologists know of several examples of such gastronomic cannibalism of strangers practiced by modern people in recent prehistory,[12] so we should not be surprised to see it with Neandertals.

There are several other, more equivocal examples of Neandertal cannibalism, the most well-known being the remains from the Croatian site of Krapina. The site was excavated over a century ago, before the development of modern techniques for extracting and preserving finds, but it is nevertheless clear that the remains included fragmentary bones of many Neandertals (perhaps upwards of forty, though here there is no reason to conclude that the remains represented a single group), and that some of these bones had cut marks and some were burned. Recent reanalysis suggests caution in attributing cannibalism to this example. Very few bones have evidence of burning, and the fracture patterns and body part representation could be the result of natural processes, especially if the individuals had received the low-investment corpse treatment attested at other sites. Neandertals returning to the cave and building fires could easily account for the few bones that have evidence of charring. Instead of cannibalism, the Krapina remains may simply reinforce an interpretation of minimal corpse treatment, in which later occupants of mortuary caves paid no heed to the scattered skeletal remains of previous occupants, even if they were relatives.

With this evidence in hand, we are in a position to revisit our list of modern mortuary features and see how Neandertals stack up.

Emotional saliency: There is no reason to suppose that Neandertals feared death any less than we do. This is an old, old response that evolved to maximize self-preservation and reproductive success. True, Neandertals regularly acted in ways we would consider incredibly dangerous (e.g., killing mammoths with thrusting spears), but this does not mean that they had no fear of death. They may have assessed risk a bit differently, which gives us some clues to personality (more on this later), but they were not fearless. Neandertals also experienced grief at the loss of a loved one. Even if we had no evidence of mortuary treatment, we could still reach this conclusion. Chimpanzee females experience an emotion that we recognize as distress when they lose an infant, and chimpanzee adults appear to grieve the loss of a close companion. Neandertals undoubtedly shared these feelings of grief, and almost certainly extended them to the loss of a spouse. Corpse treatment suggests that feelings of loss extended beyond a few hours or days. One possible motivation for hiding a corpse is to reduce the emotional pain of seeing a loved one decay. Indeed this scenario provides the minimum explanation for Neandertal corpse treatment: they did it to mitigate grief.

Social disruption: Given the small size of Neandertal face-to-face groups and communities, any loss of an adult would leave a large hole in the social fabric. If, as we have argued, the basic foraging unit was a nuclear family, then Neandertal territorial communities must have been able to incorporate the survivors into another foraging unit. The simplest way to do this would be via kinship links. But compared to modern humans, Neandertal death was less disruptive. For one, they had no significant sexual division of labor, and even a nuclear family would simply have lost one of a more or less interchangeable set of actors. Second, Neandertals had no long-range social obligations outside of their territorial community. They just had fewer social roles overall, reducing the social disruption of death.

Emotional ambivalence: It is unlikely that Neandertals felt as much ambivalence at the death of a loved one as modern humans do. Surviving family members had little to gain and much to lose. Grief would have been the dominant emotion. But there must occasionally have been ambivalence. When an adult Neandertal man or woman died, his or her spouse would become available for other men and women, and such feelings are often not trivial. But there would be no other possible advantages to the death of an adult—no wealth, and little power.

Continuance: Did Neandertals believe in some kind of continuance after death? Were there Neandertal ghosts or reincarnation? Our only line of investigation into these questions is via corpse treatment.

Ritual mitigation: Neandertals moved corpses, and often hid them with rocks or laid them in low spots, occasionally even scooping out a shallow pit. Weren't these mortuary rituals? In some way, weren't these actions a way to mitigate the emotional impact and social disruption of death? Many paleoanthropologists think so, but we urge caution. Neandertal corpse treatments were minimal, and more telling, ad hoc. As far as we can reconstruct from the evidence, Neandertals had no recurrent techniques for treating corpses. Most often they simply hid them by whatever means was easily available. Without some recurrent action patterns, it is difficult to argue for traditional, rule-governed ritual, though given the small sample, such a pattern would be hard to detect. There were no grave goods and no standard orientation. We suppose it is possible that Neandertals practiced elaborate rites that we just do not see in the paleoanthropological record. We just find it unlikely.

Transformation of the corpse: Neandertals did attend to corpses, which calls out for an explanation. We see two possibilities, one of which (grief mitigation) we have already mentioned. The second possibility is that Neandertals were actively severing the dead from the living. For some reason they found it necessary to make a physical break from the dead. Such severance is very often a component of modern mortuary treatment, and modern people provide a variety of reasons for it—fear of ghosts, sending the deceased away, and so on—most of which hinge on some notion of postmortem continuance. Perhaps Neandertals did believe that something of an individual continued after death. But if they did, it must have been a short-term continuance. Evidence from sites such as Krapina and Kebara suggest that Neandertals returned to these mortuary caves, and often lived among the disturbed bones of previously hidden corpses, whose remains were shoved aside in the same way all bones were shoved aside. If any element of the deceased was thought to have continued after death, it apparently was not thought to continue very long.

Variability: Given the ad hoc nature of Neandertal corpse treatment, we would expect a great deal of variation, which is what we find. The only commonality appears to be the minimal effort invested. Very few Neandertal corpse treatments survived more than a few years. The goal was only temporary protection.

Given what we know of Neandertal life and Neandertal corpse treatment, we think it unwise to equate Neandertal mortuary activities with modern human mortuary ceremonialism. There are clear differences in scale and content. Neandertal corpse treatment suggests actions to mitigate the emotional and social impact of death for the very short term. Large-scale rituals in which the entire community took part and which realigned and reinforced social links are just not in evidence. Nor do there appear to have been any religious or symbolic components. At most, Neandertals may have had some notion of continuance after death, but it was not one that affected the living for more than a short time.

It might be a good idea at this point for us to remind readers that Neandertals were not our lineal ancestors, at least not to a significant degree. We need not, indeed probably should not see Neandertal corpse treatment as the initial step toward modern mortuary ceremonialism. It was instead a Neandertal phenomenon that at most paralleled developments in our own past; we need not understand it in the same way. But the two separate mortuary systems may both have had roots in the behavior of *Homo heidelbergensis*. The site of Sima de los Huesos at Atapuerca in Spain (about 400,000 years old) consists of a vertical shaft inside a limestone cavern, filled with the bones of at least thirty-two *Homo heidelbergensis* and other animals.[13] The bones are jumbled, and there are no articulated skeletons, yet the sheer number of individuals, and their location well inside a twisting cave chamber, make it difficult, though not impossible, to account for them by natural processes alone. Many paleoanthropologists think that someone deposited the corpses, or just the bones, in the shaft. There was even a nice handaxe found among the bones. As always, we must take care not to read modern behavior into this evidence. We see it as possible evidence that *Homo heidelbergensis* moved corpses to a protected place. If true, it was a treatment very like the one we have just described for Neandertals. Once again Neandertals come across as lineal descendants of an earlier European way of life.

By 160,000 years ago, and a continent away, our ancestors may already have been treating the dead differently. At Herto in Ethiopia Tim White and colleagues discovered the skulls of two early *Homo sapiens sapiens*.[14] The shapes of the crania and faces were clearly not those of Neandertals. But what is most provocative is evidence that someone altered the heads after death. Cut marks clearly indicate that the heads had been defleshed by stone tools. This time it was not for cannibalism. One of the skulls had polish marks that resulted from friction with softer material, possibly leather. Someone had carried this skull around in a bag. This is different. All of the evidence of Neandertal mortuary treatment, however meager, points toward physical separation and severance from the living. If the Herto skull treatments were somehow typical, and it is dangerous to generalize from such single examples, then *Homo sapiens sapiens* were doing something rather different, perhaps emphasizing continuity as opposed to severance.

Fire

Fire is an oddly neglected topic of study in recent anthropology, despite its immense importance in modern life. Control of fire and its heat are at the core of modern industrial economies—indeed harnessing the heat of burning coal powered the Industrial Revolution—and coping with the by-products of burning has become one of the banes of modern existence. Without cooked foods, especially cooked cereals, there would be no way to feed the earth's 6 billion people (or make beer). And without fire as a source of heat our tropical bodies would be unable to survive in Europe or North America or most of Asia. Yet a comprehensive anthropological study of the evolution of pyrotechnology has yet to be written, and this is a puzzle. Paleoanthropological treatments of fire rarely extend beyond documenting its presence or absence. Neandertals did use fire, often and effectively, but the range of uses differed slightly from ours, and this difference points to possible differences in Neandertal minds.

Imagine for a moment a European family of 600 years ago sitting near their hearth. What are they doing? Keeping warm, to be sure, and perhaps eating cooked food, but also talking. This talk is likely to have been mostly gossip, but there would also have been stories, such as the oral tradition of *Hansel and Gretal* and perhaps *Rumpelstiltskin*. Now think of the great feast hall of Hrothgar in *Beowulf*, where deeds of heroes were sung, or even of a troop of Girl Scouts in the evening after a day's hike. All sit around or near a fire, which becomes more than a warm spot. It becomes the visual focus of the interaction. Participants stare at the fire or coals, and their minds turn inward to the events and characters of the story. It is a powerful experience, especially for the young. And fire itself is often an element in the stories: Prometheus giving fire to people, or the witch's oven in *Hansel and Gretel*. Of course, this experience is not limited to Europe or European folk tales. It is a universal component of the human experience. Warmth and cooking may well have been the first roles for fire in human evolution, but somewhere and at some time it acquired the social, symbolic role it has for us today.

Hominins harnessed fire long before Neandertals appeared on the evolutionary stage. Exactly who were the first pyrotechnicians remains controversial, largely because the most direct evidence of fire use, charred wood or charcoal, rarely preserves longer than a few hundred thousand years. The oldest, widely accepted direct evidence for use of fire comes from the Israeli site of Gesher Benot Ya'aqov, where sediment preserved the remains of burnt seeds and wood, all associated with stone tools. The site is about 790,000 years old, early enough that the pyrotechnicians must have been *Homo erectus*.[15] More controversial evidence pushes back use of fire to perhaps 1.2 million years ago. The South African cave site of Swartkrans yielded the remains of burnt, butchered bones.[16] There is no charcoal and no possible hearth features or heat-altered sediment, and some paleoanthropologists have suggested that the discoloration of the bones either had a chemical source or was somehow the result of natural burning.

Indirect evidence supports the use of fire at this earlier date. Paleoanthropologists now generally agree that the increase in brain size that accompanied the evolution of early members of the genus *Homo* was enabled or powered by an increase in meat consumption. Meat provides the necessary calories and protein to support expanded neural tissue, but only if it can be digested efficiently. But, as the Harvard primatologist Richard Wrangham argues, raw meat is tough and hard to chew and digest unless it has been pounded or cooked. Pounding meat with stones may have been our ancestors' first solution to the problem, but by the time (and brain size) of *Homo erectus*, cooking may well have been required.[17]

By the time of Neandertals, fire use and cooking had long been established components of hominin life. Neandertals could not have survived in the cold of glacial Europe, nor consumed the massive quantity of meat that was their primary diet, without fire. They clearly knew how to keep and maintain it. We suspect that they also knew how to make it through friction, or perhaps even by creating sparks using meteoric iron to strike flint. No one, to our knowledge, has explored the cognitive implications of fire making and use, but it would appear to be well within the range of the expert procedures well documented for Neandertals.

Neandertal Hearths

Among the most meticulously documented Neandertal hearths are those excavated at Abric Romani, a rockshelter located about 50 kilometers northwest of Barcelona.[18] The site itself has over 17 vertical meters of deposits dating between 40,000 and 70,000 years old. Several levels in this sequence contain multiple "combustion activity areas" whose careful excavation provides arguably our best glimpse at Neandertal pyrotechnology.

Figure 5-3 is a map of the burned areas of one level at Abric Romani. At first glance there appear to be at least two sizes of fire: small ones, labeled 1 to 5, and larger ones, such as 8, 13, and 16. But careful excavation revealed that the larger burned areas resulted from smaller fires being rebuilt several times. The small combustion areas are the remains of simple, flat hearths built on the floor of the rockshelter. The fires were shallow and small, covering less than 3 square feet (0.28 square meters). The condition of surrounding sediment indicates low-temperature burning, under 300° C. The larger, reused hearths are associated with single episodes of activity, either someone knapping a single core near hearth 8, or butchering deer between hearths 9 and 12. Some of the small areas, 1 to 5, were single, small fires with little associated activity refuse. These may have been sleeping areas. The activity debris was often distributed on only one side of the fire in small patches appropriate for one person and one activity. All in all, the picture of fire use at Abric Romani is of small fires,

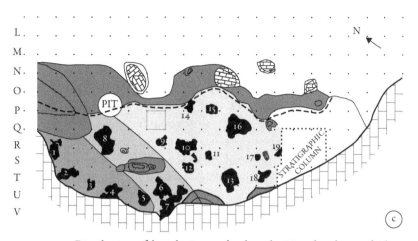

FIGURE 5-3 Distribution of hearths in one level at the Neandertal site of Abric Romani (after Vallverdu et al.[18]).

occasionally rebuilt, that did not burn very long or very hot. Neandertals used them for warmth and cooking, in activity areas and sleeping areas.

Most Neandertal hearths resembled those at Abric Romani. The only good example of something a bit more organized comes from Kebara Cave in Israel.[8] Figure 5-4 is a photo of the stratigraphy from the central part of the cave. The dark and light bands consist of ash and burnt sediment from

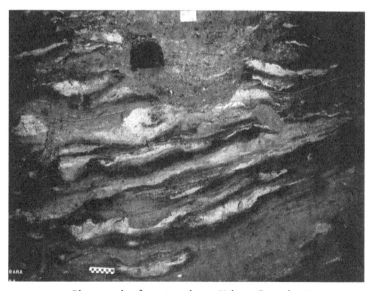

FIGURE 5-4 Photograph of stratigraphy at Kebara Cave showing superimposed hearths (after Bar Yosef et al.[8]).

several hearths superimposed one atop another. Most of these hearths were similar to those at Abric Romani: flat, unprepared fires, 20 to 80 centimeters in diameter. At some of the hearths it appears that Neandertals had intentionally spread the ashes. A few of the hearths were larger than those at Abric Romani, including one over 1 meter in diameter and 30 centimeters thick. It is not clear if this feature resulted from a continuous long burn or multiple rebuilds of the same fire, but it was certainly larger than was typical for most Neandertal fires. There is also some evidence that Neandertals cleared ash from some of the fires and dumped it elsewhere in the cave.

Hearths similar to the Neandertal hearths just described are typical of prehistoric hunter-gatherer groups all over the world—individuals building fires for cooking and warmth in an ad hoc manner. This pragmatic use of pyrotechnology has been a component of human adaptations for at least several hundred thousand years, perhaps longer. Like their ancestors, Neandertals used fire as a practical component of their everyday lives. Neandertals were neither precocious nor backward in this regard. Fire helped keep them warm in cold conditions and transformed their food into a more easily digestible form. They may even have used fire to alter the flaking characteristics of stone. As we have seen in many kinds of activities, it was not the Neandertals who were unusual; it was *Homo sapiens sapiens*.

Most of the hearths archaeologists can attribute to prehistoric modern humans resemble those of Neandertals, but not all. About 23,000 years ago at the rockshelter of Abri Pataud in southwestern France, modern people scooped out shallow pits and lined them with river cobbles.[19] There were also larger hearths, over a meter in diameter. At the slightly earlier Czech site of Dolni Vestonice are the remains of large hearths over 2 meters in diameter with deposits over 40 centimeters (over a foot) thick, attesting to continued, intensive use.[20] Moreover some of the Dolni Vestonice fires were associated with shattered clay figurines, and clay pellets apparently made to expand rapidly and explode in the fire. Dolni Vestonice even boasts two kilns, clay structures built to produce very high temperature burning for firing the clay figurines. Fire had become much more than a practical tool for warmth and cooking. It had become a focus of social activity.

Imagine for a moment the activities surrounding a hearth at Dolni Vestonice. The shattered figurines and clay pellets point to social interaction, not just shared warmth and cooking. Someone placed these things in the fire and watched them shatter and explode. We suppose it might just have been preadolescent boys on a lark with Stone Age fire crackers, but there are really too many shattered pieces for this to be the most likely explanation. Following Clive Gamble, we think someone was telling stories, using figurines and clay pellets as props in the performance. Hearths had become a focus of social life. Long after the cooking was done, people sat around the fires and talked, recited myths, and performed rituals. This is a familiar use of fire in the twenty-first century, and has been for at least 25,000 years. It is this kind of fire use that has never been found with Neandertals. It marks an important social difference between Neandertals and us, and also a possible difference in the nature of mind and language.

The psychologist Matt Rossano thinks he knows why modern humans built larger, longer burning, and more intense fires: they used them in their rituals.[21] By ritual Rossano is referring to something both more inclusive than the common notion of religious ritual, and also something more narrowly defined. Rituals are simply activities performed according to sets of procedures that are rule-governed and formalized. From this point of view any repetitive sequence of action that one performs in an invariant way would qualify, so a major league batter tapping the plate with his bat three times, taking two slow swings, and then setting his stance is following a ritual. Why does he do it? To "get in the zone" and to "focus" are common explanations. Cognitively both imply focusing attention, which is an aid to performance. Such rituals are important components of expert performances of all kinds, including the ones so common in Neandertal life. So what was so different about the rituals of *Homo sapiens sapiens*?

Modern humans intensified ritual performance in at least two ways. First, they enhanced the quality of focused attention through consciousness-altering meditation (sometimes aided by mind-altering chemicals). Second, they invented group rituals in which many individuals together coordinated attention. Both had short-term and long-term effects on the brain itself.

Research using electroencephalography (EEG) and functional magnetic resonance imaging has clearly documented that meditation changes brain activity. In one EEG study experienced meditators were able to coordinate neural activity across the frontal and parietal lobes of the brain to a degree that was far beyond the abilities of normal individuals. There is even some evidence for permanent changes induced by long experience with meditation. Some meditators have significantly thicker regions of the prefrontal cortex than is typical. This is perhaps not surprising; neuroscientists have long known that the brain has some plasticity, with heavily used areas expanding slightly (e.g., in string musicians the area of the motor cortex devoted to fingertips is larger than that of nonmusicians). So what areas of the brain become enlarged in meditators? Primarily areas of the prefrontal cortex devoted to working memory and focused attention. Meditation is a kind of ritualized thinking; by following a series of action and thought patterns the meditator marshals his or her neural resources to focus on a narrow range of stimuli, either external or internal. The goal is an altered state of consciousness. It takes practice because it is not easy.

Shared rituals are a bit different. The goal here is for a group to achieve some level of altered awareness. Most modern religions have group performances in which some or all participants strive for ecstasy, a feeling of disembodied joy, or terror. Speaking in tongues, being "saved," and experiencing miraculous healing are just a few examples. Psychologists and physicians have long known that healing rituals can have very real psychological and physical effects. Much of the efficacy of such ritual depends on a participant's hypnotizability, the ability to respond to suggestion induced by ritual performance, and more often than not such suggestibility is actively manipulated by a religious practitioner—a shaman, a priest, or even a charismatic preacher. The cognitive and neurological effects are similar to those of meditation, and like meditation, practice improves performance. The states of altered consciousness are highly valued by most participants. Their evolutionary significance is not as obvious. Rossano favors group selection explanations in which ritual enhances group solidarity and success, but this need not concern us for the moment. The question at hand concerns fire. What role, if any, did fire play in the emergence of these consciousness-altering activities?

The answer to this question could be quite mundane: group rituals often occurred after dark, and the only light available came from fires or, in cold places, the only communal warmth came from fire. But somehow this mundane explanation doesn't seem quite adequate. Why is fire common in rituals performed in daylight, or on warm days, or in the tropics? Fire is strangely alive. It flickers and dances, and the coals seem to throb with vitality. We suspect that fire helps individuals and groups focus attention. It is visually dynamic. It constantly changes, but it does not move, and the more suggestible someone is, the more effective it is. This is why people invested effort to build fireplaces that would allow long-burning fires. *Homo sapiens sapiens* used them for more than cooking; they used them to help create a spiritual life. Neandertals never appear to have done this. It is one of the most profound differences between them and us.

Artless Dodgers

Most people have heard of cave paintings, naturalistic images of animals painted on cave walls deep underground. These evocative depictions often hold the place of honor as the earliest art in art history courses. They are one of the few bits of Palaeolithic evidence to have become a component of the general knowledge base of Western culture. Neandertals didn't paint them; modern humans did. This single difference—no art from Neandertals, art from moderns—is partly responsible for the uncouth, primitive picture of Neandertals most people carry in their heads. It is a bum rap, to use a colloquial phrase, not because Neandertals painted caves, but because very, very few modern humans, today and in the past, crawl down into dark caverns to paint images on wet cave walls. The Franco-Cantabrian tradition, to use a more formal term for cave painting, was a phenomenon limited to a narrow region of western Europe.[22] It should not be seen as a necessary feature of modern behavior, even though modern people were undoubtedly responsible. But it is fair to ask if Neandertals produced anything of a similar nature, anything that might be counted as depiction, or decoration, or ornamentation, because with very few exceptions all modern humans do at least some of these things.

Neandertals produced no depictions. There is nothing in the extensive archaeological record that could in any way be construed as a Neandertal's representation of an animal, or a person, or indeed of anything. This lack of images is not perhaps a serious deficit; many modern human societies lack a tradition of representational images, and some even prohibit them outright. But virtually all contemporary societies have traditions of decoration, altering the appearance of objects and tools and of human bodies themselves. The reasons for decoration are varied; some decorations mark ownership, some mark social status and group membership, some are done to attract the opposite sex, and some are just aesthetic upgrades. Whatever the reason, decoration is virtually ubiquitous in the material culture of modern humans. And decorations can leave archaeological traces in the guise of decorated objects themselves and the media used for decoration.

The Neandertal archaeological record does offer several provocative items.[23] In some sites archaeologists have recovered rock crystals that Neandertals apparently picked up and carried home. At the Hungarian site of Tata archaeologists found a round river cobble that had a straight crack running from one edge to the other. A Neandertal had engraved a second line perpendicular to the crack. From two sites in Spain, Cueva de los Aviones and Cueva Antón, there are perforated marine shells most likely used as pendants.[24] Such artifactual examples suggest an attraction to patterns and perhaps a concern for appearance, but there are so few of them that it is dangerous to generalize. (Indeed until the Cueva de los Aviones pendants were found there were no unambiguous examples of Neandertal ornaments in the entire archaeological record.) Use of mineral pigments was far more common. These are minerals such as hematite and manganese dioxide that produce a distinct colored mark when scraped across a surface or which can be ground into colored powder. The most common is known as ochre (hematite), which comes in colors from red through yellow and orange. It is common in prehistoric sites around the world, and its earliest use extends back as far as 300,000 years ago in Africa, to the time of *Homo heidelbergensis*. Neandertals also used ochre, but manganese dioxide was even more common; pieces of manganese have been found in more than forty European Neandertal sites. Most of the pieces have evidence of scraping to produce a black powder, and some were ground into pointed shapes, perhaps to use as a kind of pencil.

What were Neandertals doing with the pigments? There are two possibilities, one mundane and one provocative. The mundane explanation is that the powder was a component in some Neandertal technology. In southern Africa modern human hunter-gatherers mixed ochre powder with plant gum and beeswax to produce the glue that they used to attach points and barbs to spear shafts. The residue remains on some of these tools, which are over 70,000 years old. Perhaps Neandertals used manganese powder similarly as a binding agent for their glues, but there is no evidence for such use. The more provocative explanation is that Neandertals used the manganese powder to color something, perhaps their bodies, an interpretation given more power by the rare examples of pendants.

From a cognitive perspective, the second explanation is perhaps the more interesting. Our knowledge of Neandertal technology already yields a picture of expertise along with multicomponent tools, and adding powdered bonding material to the picture does not change this. But if Neandertals painted their bodies and used ornaments it would have interesting implications. You do not generally alter your appearance for yourself; you do it to change how you appear to others. At a minimum this requires a Theory of Mind (I know that you see me), but it also suggests conscious effort to change how you relate to someone else. It needn't have been symbolism; black or red need not have stood for anything, but it does suggest some intentional marking or manipulation of role. We just do not know. However we see it, it is important to note that for Neandertals *and* modern humans it was an old behavior that can be traced back to *Homo heidelbergensis*, who was the first to use pigments not just in Africa, but in Europe as well. It is something Neandertals and modern humans shared. For Neandertals it went no further. For modern humans it may have been the basis for a variety of other forms of decoration, from geometric markings to beads. Once again Neandertals appear conservative and modern humans more innovative.

Symbols or Not?

We have dwelt on the evidence for burial, social use of fire, and ornaments and body coloring because it has often been used to argue for a shared humanity between Neandertals and modern humans. But this interpretation does not hold up to close scrutiny. Yes, in each case there are similarities. Neandertals did devote some care to corpses, did use fire for warmth and cooking, and did use pigments and, occasionally at least, personal ornaments. But the differences are striking. Neandertal corpse treatment was minimal, they did not use fire for expanded social interaction, and they did not have elaborate decorative traditions.

From our cognitive perspective, the tenuous evidence we have for a Neandertal symbolic life is consistent with the picture of small-scale, embodied social cognition presented in Chapter 4. Treating corpses to mitigate grief and coloring bodies both fit easily into a way of life based on face-to-face interaction, gesturing, touching, and emotive vocalization. The lack of intense fire suggests that Neandertals did not have any kind of narrative tradition or symbolic ritual life. As in other regards, Neandertal thinking appears to have been concrete, in the here and now. They could think into the future, else why bother with corpses, but this thinking was limited in scope. Their tendency to rely on expert systems meant that their cognitive resources were devoted to real solutions, to tangible, real-world problems.

Neandertal life was not immersed in symbols. They may have had some components of symbolic culture; for example, use of pigments might have played a conscious indexical role. But we have very little evidence for even this. The Neandertal evidence for symbolism fits most easily into an embodied social model in which personal appearance was a component. It also reinforces our assessment that Neandertals had a working Theory of Mind ("I know that you know that I know"). But this alone hardly constitutes immersion in symbols.

Of course, we have thus far avoided a very important piece in the puzzle of Neandertal symbolic life: language. It is the focus of the next chapter.

Further Reading

1. Danesi, M. 2008. *Of Cigarettes, High Heels, and Other Interesting Things.* New York: Palgrave Macmillan.

2. Turner, V. 1967. *The Forest of Symbols.* Ithaca, N.Y.: Cornell University Press.

3. Wynn, T., F. L. Coolidge, and M. Bright. 2009. Hohlenstein-Stadel and the evolution of human conceptual thought. *Cambridge Archaeological Journal* 19(1): 73–83.

4. Goody, J. 1962. *Death, Property and the Ancestors: A Study of the Mortuary Customs of the LoDagaa of West Africa.* Stanford: Stanford University Press.

5. van Gennep, A. 1960. *The Rites of Passage.* London: Routledge & Kegan Paul.

6. Szyniewicz, S. 1990. Philosophy of the corpse: Modes of disposal and their cultural corrlates. In *The Life of Symbols*, eds. M. L. Foster and L. J. Botscharow. Boulder, CO: Westview Press.

7. Renfrew, C., ed. 1990. *The Prehistory of Orkney.* Edinburgh: Edinburgh University Press.

8. Bar Yosef, O., et al. 1992. The excavations in Kebara Cave, Mt. Carmel. *Current Anthropology* 33(5): 497–550.

9. Gargett, R. 1989. Grave shortcomings: The evidence for Neandertal burial. *Current Anthropology* 30:157–190.

10. Gargett, R. 1999. Middle Palaeolithic burial is not a dead issue: The view from Qafzeh, Saint-Cesaire, Kebara, Amud, and Dederiyeh. *Journal of Human Evolution* 37:27–90.

11. Defleur, A., et al. 1999. Neanderthal cannibalism at Moula-Guercy, Ardeche, France. *Science* 286(5437): 128–131.

12. White, T. 2001. Once we were cannibals. *Scientific American* 285(2): 58–65.

13. Carbonell, E., and M. Mosquera. 2006. The emergence of a symbolic behaviour: The sepulchral pit of Sima de los Huesos, Sierra de Atapuerca, Burgos, Spain. *Comptes Rendus Palevol* 5:155–160.

14. White, T., B. Asfaw, and D. DeGusta. 2003. Pleistocene *Homo sapiens* from Middle Awash, Ethiopia. *Nature* 423:742–747.

15. Alperson-Afil, N., D. Richter, and N. Goren-Inbar. 2007. Phantom hearths and the use of fire at Gesher Benot Ya'aqov, Israel. *PaleoAnthropology* :1–15.

16. Brain, C. K., and A. Sillen. 1988. Evidence from the Swartkrans cave for the earliest use of fire. *Nature* 336:464–466.

17. Wrangham, R. W. 2009. *Catching Fire: How Cooking Made Us Human.* New York: Basic Books.

18. Vallverdu, J., et al. 2010. Sleeping activity area within the site structure of archaic human groups: Evidence from Abric Romani Level N combustion activity areas. *Current Anthropology* 51(1): 137–145.

19. Movius, H., ed. 1975. *Excavation of the Abri Pataud, Les Eyzies (Dordogne).* Cambridge, Mass.: Peabody Museum of Archaeology and Ethnology.

20. Gamble, C. 1999. *The Palaeolithic Societies of Europe.* Cambridge, U.K.: Cambridge University Press.

21. Rossano, M. 2007. Did meditating make us human? *Cambridge Archaeological Journal* 17(1): 47–58.

22. Ucko, P., and A. Rosenfeld. 1967. *Palaeolithic Cave Art.* New York: McGraw-Hill.

23. d'Errico, F., et al. 2003. Archaeological evidence for the emergence of language, symbolism, and music: An alternative multidisciplinary perspective. *Journal of World Prehistory* 17(1): 1–70.

24. Zilhao, J., et al. 2010. Symbolic use of marine shells and mineral pigments by Iberian Neandertals. *Proceedings of the National Academy of Sciences* 107(3): 1023–1028.

6

Speaking of Tongues

WHEN THE TWO of us began working together more than a decade ago, the archaeologist of the pair (TW) made two explicit stipulations: he did not "do" Neandertals, and he did not "do" language evolution. These aversions were not due to lack of background (indeed he had started his undergraduate training in anthropological linguistics); they were due to a reluctance to get bogged down in topics that generated so much heat but so little light. Clearly he has not been able to avoid Neandertals. And despite continued reservations, he now agrees that for the purposes of this book it is necessary to address language evolution. Language remains the one human characteristic that is truly unique; no other animal—not chimpanzees, not orcas, not parrots—uses a form of communication even vaguely similar to language. Many scholars even argue that it is the pivotal feature that makes us human. If we want to compare Neandertals to humans, indeed if want to document how human they were, we must try to document Neandertal language. Did they have language or not? If not, did they communicate in some way very like modern language? These are very difficult questions to answer, maybe even impossible. But before we discourage you about our prospects, we need to define language. Even here, in trying to accomplish this seemingly simple task, we will undoubtedly annoy many who study language evolution, but there is no help for it. Language is just far more subtle and complex than most people realize.

Language for the Uninitiated

There are three major components of language proper, and several related cognitive components that are tied to its delivery.

Words

"KILL LA COTTE MAMMOTH YOU I"

Each of these words is a symbol; it refers to something. When you hear the word "mammoth," or see it written, your mind calls up an image or idea of the creature. Exactly how the brain does this is an interesting story in itself, but for the current discussion we will just accept that it does. Using words, you can call up things and ideas that are not actually available to immediate perception, so their utility in communication should be obvious. Of course, the words in our example do not refer to the same kinds of things. "Mammoth" and "La Cotte" are pretty straightforward—one is a kind of animal and one a specific place—so it is fairly simple to assign them each a label (assuming that you have experience with them). "Kill" is not as specific about its reference, and indeed assumes that the speaker and hearer know about death. "You" and "I" do not actually have a permanent reference. My saying "I" will not refer to the same thing as your saying "I." These distinctions between kinds of words mean that words just strung together, as in our example, carry a lot of communicative potential.

We are not the only animals who can do this. Vervet monkeys use different calls to warn about different predators.[1] More famously, several great apes have been taught to use words (usually in the form of visual patterns called lexigrams) to communicate with humans. They learn the difference between nouns and verbs, and even less concrete references such as "happy" and "sad." The most adept ape "signers" acquire a vocabulary of hundreds of words. They can even string words together in proper order, understanding that "Kanzi give Bob drink" is very different from "Bob give Kanzi drink." Word order is not a feature of words themselves; it is feature of syntax, and its use by apes indicates that they can learn to understand and use this second component of language. There are some words, however, that apes have a very hard time with, words such as "the" and "a."[2] The reference of these words is very peculiar—certainly from an ape's perspective. In a sense, they don't "stand for" anything, but they do affect meaning, primarily through the rules of a language.

Rules

"'Twas brillig and the slithy toves did gyre and gimble in the wabe."

Most English speakers have no difficulty recognizing this example as an acceptable English sentence and are even able to extract meaning from it, despite its being nonsense. Indeed its undoubted charm derives largely from the absence of any clear reference performed by the key words. So how can it be English, or even language? The answer is "by following English rules." Let's look at the major words—the adjectives, nouns, and verbs. First, "slithy." We know that it is an acceptable English word because it follows the rules in English for stringing sounds together; "s" and "l" followed by a vowel is a common grouping in English (e.g., slug, slippery), as are "th" and "y." The word "mbwa" wouldn't work, however, because "m," "b," and "w" never go together at the beginning of an English word (they do in Swahili). We also know that "slithy" is an adjective describing "toves" because of its position in the sentence and its terminal "y," which often ends adjectives. Similarly "brillig," "gyre," "gimble," and "wabe" all follow the English rules for making words, even though they have no known reference. (We will conveniently ignore the very real fact that certain sound combinations in English do tend to evoke certain feelings, even a kind of vague reference.) The lesser words in the sentence—"twas," "and," "the," "did," "in"—are all words that, combined with word order, make the sentence grammatically English. These guide the reader (or hearer) through the words. What Lewis Carroll did here, in the first line of "The Jabberwocky," was produce a perfectly intelligible bit of nonsense by following the rules of English without any specific reference. In other words, the rules of language are powerful enough to generate meaning without concrete reference of any sort. But of course syntax, the rules for stringing words together, typically does organize words, and when the rules of syntax organize the 30,000-plus words in a typical speaker's vocabulary the result is an incredibly powerful tool for expression. Every normal human speaker is able to utter a sentence no one has ever uttered before—"The jazz bassoonist played Bach in a funky style"—with little or no effort, something linguists refer to with the uninspiring term "productivity." Every normal speaker is able to describe everything in his or her world, and even in worlds just imagined.

Usage

If words and syntax weren't subtle and powerful enough, all languages have various "modes" of speech, which are the kinds of things that the language can be used to express. Some are simple and straightforward, such as exclamations like "S***!" which pretty clearly express an immediate emotional state (and are therefore much like any other primate vocalization in that their primary reference is an emotional state). Languages can pose questions ("Where is Bruce?") and make commands ("Give me food"), which are fairly simple manipulations of information. And there are more subtle things. Languages can call attention to things ("This book"), and can also express hypotheticals ("Were I king"). All of these differences can be indicated through different rules (e.g., combining "I" with "were" in English to produce the subjunctive), or different words, or both, but the modes of speech also reflect the different kinds of things that language has evolved to do. Even more subtly, every native speaker of a language is able to adjust word choice and rule choice to better fit myriad social interactions. The language you use when speaking to a police officer will be very different from the language you used minutes earlier in the bar with your friends. Linguists refer to this as "code switching," and we all do it, largely unconsciously. We each use many slightly different languages.

IN SUM, LANGUAGE is a sophisticated and subtle form of communication. It is central to our modern lives; we could not exist without it. But what about Neandertals? Did they too live their lives through the vehicle of language? If so, was it different from ours? These are important questions, but they are very hard to answer.

There is, as you might imagine, a serious academic interest in the evolution of language,[3] and in Neandertal language in particular. It has produced a voluminous, often interesting, but rarely conclusive literature. There are two major roadblocks to documenting the evolution of language: first, no living animal other than humans uses a form of communication even remotely like language, and second, language itself does not preserve in either fossils or archaeological remains, at least not until the advent of writing. This means that the two major methods of evolutionary science, paleontology and the comparative method, have nothing directly to offer. We cannot study animals that use a pre- or proto-language, and we have no examples of ancient pre- or proto-languages. We are left with only two alternatives: documenting the evolution of anatomies or behavior related to language in some way, and arguing about evolution from the nature of language itself. Neither has provided a clear picture of Neandertal language.

Use of spoken language has almost certainly selected for specific anatomical and neurological features whose evolution could, in principle, cast light on the evolution of language itself. It is very likely, for example, that speech has selected for efficient anatomies for producing a rich range of sounds and a sensitive system for detecting sound.[4,5] Unfortunately none of the anatomies required—vocal chords, throats, tongues, lips, cochlea, and so on—preserve in fossils. Paleoanthropologists have made a serious attempt to describe changes in the bone structures that support these speech organs (e.g., the hyoid bone), but, to be honest, the result has been disappointing. We cannot pinpoint the first fossil hominin to use speech, and, more relevant to our current task, we cannot describe how Neandertal speech or hearing might have differed from ours. Given the differences between Neandertal faces and ours, we suspect that Neandertal voices sounded a bit different, and the range of consonants and vowels they could generate might have been different, but nothing we know about Neandertal anatomy would seriously limit their capacity for speech. Evidence from fossil endocasts of Neandertal skulls is no better. We know that modern brains control speech production using a region of the lower left frontal lobe known as Broca's area and that, compared to other primates, our Broca's area is enlarged.[6] So when we see evidence of enlargement of Broca's area on endocasts of fossil skulls we have a smoking gun for speech, right? Again it is not so easy. In nonhuman primates this area of the brain controls guided reaching with the hands, the kind of motor action that is also used to manipulate tools. And the first evidence for expanded Broca's area in fossil hominin endocasts coincides with the first appearance of stone tools. So was it speech or tool use? We don't know. On the positive side, evidence conclusively identifies an expanded Broca's area for all post–*Homo erectus* hominins, including Neandertals. This suggests that Neandertals had speech, but does not tell us how similar it might have been to ours.

Stone tools also offer little help. There has been an understandable hope on the part of archaeologists that regularities in stone tool manufacture might reflect rule-governed thinking, and that such rules might be evidence for the presence of grammatical rules. If true, the millions of stone tools archaeologists have recovered spanning 2.5 million years might have an interesting tale to tell. Unfortunately, as we discussed in Chapter 3, technological cognition is largely separate from language, and the regularities apparent in stone technologies derive from a very different way of thinking than that of syntax. The difference in how technology and language are learned is especially informative; young children learn syntax effortlessly if they are raised in a linguistic environment, but children and adults require years of dedicated practice to learn a complex technology. However, learning does provide one possible avenue for getting from stone tools to language. Some technologies, Levallois for example, may have required some verbal instruction. This is a reasonable argument (though it would be stronger if the proponents could specify just what required verbal instruction) and confirms the suspicion derived from endocasts that Neandertals may have had some form of verbal communication that conveyed factual knowledge at least.

Ochre and manganese dioxide use are also provocative.[8] If Neandertals colored their bodies and occasionally attached ornaments to change how they appeared to others, this would suggest not only Theory of Mind, but also a kind of symbolism. Perhaps the colors or patterns (but remember we have no actual examples of the latter) stood for something—ferocity, or leadership, or sexual status. This is akin to the symbolism in words in that something would be standing for something else (with the something else not, perhaps, being in view).[7] But this is a precarious chain of reasoning, from a lump of scraped manganese to words, and the reader is forgiven for feeling a tad skeptical. The use of mineral pigments suggests that Neandertals may have used colors as a kind of visual information, and this is similar to word use. But it does not require modern language.

For the general study of language evolution, knowledge of other species' communication is critically important. It reveals the range of communication systems that have evolved, and also the circumstances of that evolution. Science now knows a great deal about communication in bees, birds, whales, primates, and many other animals. And we now know, as mentioned earlier, that no animal uses a communication system even remotely similar to language. But our understanding of animal communication, especially that of monkeys and apes, does provide valuable raw material for thinking about language evolution. This is especially pertinent to understanding early hominins. But it is less helpful for Neandertals, who were contemporaries of *Homo sapiens sapiens* and as distant from apes and monkeys in an evolutionary sense as we are ourselves.

We are left with language itself. Linguists have approached the evolution of language by studying how children acquire it, how languages differ from one another, and how new languages such as pidgins and creoles develop. From such studies we know that human children come endowed with a "language acquisition" ability that is part of their normal genetic makeup. If raised in a language-using environment, normal children acquire it effortlessly. This must have evolved at some point in human evolution. Linguists also know that no human language is either more or less complex than any other. Recently there has been much written about a South American people, the Pirahã, who have been reported to lack several syntactical features long thought to be true of all human languages (recursion, the ability to embed an element in another example of the element, as in "John said that Bob said...," is the most salient example).[9] Even if confirmed, Pirahã language would present no necessary implications for language evolution. The Pirahã are surrounded by peoples who possess these universals, and they are related to those peoples. All native South Americans are recent arrivals, in an evolutionary sense (less than 15,000 years), and their ancestors carried with them a language with all of the expected syntactical complexities of modern language (which all but the Pirahã still share). The odd features of Pirahã must be recent developments.

Pidgins have held linguists' attention because they form initially when two languages come in contact and a functional communication system must form for interaction. Initially at least, pidgins tend to be very simple grammatically. But once children start learning them the pidgins rapidly change into creoles, which are fully complex languages. Using knowledge of pidgins and creoles, the linguist Derek Bickerton has developed a model of what a proto-language might have been like.[10,11] In essence, Bickerton's proto-language consists of words without the organizing rules of syntax. Thus without other cues, the words "KILL LA COTTE MAMMOTH YOU I" would carry the same information as "MAMMOTH KILL LA COTTE I YOU." Obviously the more words uttered in proto-language, the more chance for confusion, but context and other cues can reduce confusion considerably, and a proto-language could be very effective.

Neandertal Vocal Communication

Neandertal communication must have been different from modern language. To repeat a point made often in this book, Neandertals were *not* a stage of evolution that preceded modern humans. They were a distinct population that had a separate evolutionary history for several hundred thousand years, during which time they evolved a number of derived characteristics *not* shared with *Homo sapiens sapiens*. At the same time, a continent away, our ancestors were evolving as well. Undoubtedly both Neandertals and *Homo sapiens sapiens* continued to share many characteristics that each retained from their common ancestor, including characteristics of communication. To put it another way, the only features that we can confidently assign to both Neandertals and *Homo sapiens sapiens* are features inherited from *Homo heidelbergensis*. If *Homo heidelbergensis* communicated via modern style words and modern syntax, then we can safely attribute these to Neandertals as well. Most scholars find this highly unlikely, largely because *Homo heidelbergensis* brains were slightly smaller than ours and smaller than Neandertals', but also because the archaeological record of *Homo heidelbergensis* is much less "modern" than either ours or Neandertals'. Thus we must conclude that Neandertal communication evolved along its own path, and that this path may have been quite different from the one followed by our ancestors. The result must have been a difference far greater than the difference between Chinese and English, or indeed between any pair of human languages. Specifying just how Neandertal communication differed from ours may be impossible, at least at our current level of understanding. But we can attempt to set out general features of Neandertal communication based on what we know from the comparative, fossil, and archaeological records.

As we have tried to show in previous chapters, the paleoanthropological record of Neandertals suggests that they relied heavily on two styles of thinking: expert cognition and embodied social cognition. These at least are the cognitive styles that best encompass what we know of Neandertal daily life. And they do carry implications for communication. Neandertals were expert stone knappers and relied on detailed knowledge of landscape and a large body of hunting tactics. It is possible that all of this knowledge existed as *non-linguistic* motor procedures learned through observation, failure, and repetition. We just find this unlikely. If an experienced knapper could focus the attention of a novice using words, it would be easier to learn Levallois. Even more useful would be labels for features of the landscape, and perhaps even routes, enabling Neandertal hunters to refer to any location in their territory. Such labels would almost have been required if widely dispersed foraging groups needed to congregate at certain places (e.g., La Cotte). Most critical of all, in a natural selection sense, would be an ability to indicate a hunting tactic prior to execution. These labels must have been words of some kind. We suspect that Neandertal words were always embedded in a rich social and environmental context that included gesturing (e.g., pointing) and emotionally laden tones of voice, much as most human vocal communication is similarly embedded, a feature of communication probably inherited from *Homo heidelbergensis.*

At the risk of crawling even farther out on a limb than the two of us usually go, we make the following suggestions about Neandertal communication:

1. Neandertals had speech. Their expanded Broca's area in the brain and their possession of a human FOXP2 gene both suggest this. Neandertal speech was probably based on a large (perhaps huge) vocabulary: words for places, routes, techniques, individuals, and emotions. We have shown that Neandertal expertise was largely based on long-term memory. Much of this store of information was in the form of procedures, but we suspect an equally large part of this "how to" information existed as verbal knowledge in the form of words.

2. Many of these words existed in stock sayings, also held in long-term memory, much like the idioms and adages in modern language (the "If it ain't broke, don't fix it" variety).

3. Speech depended heavily on environmental and social information to

disambiguate word clusters. Clues from context can be very effective. American readers may remember a television commercial in which four young men riding in a car each used only a single word, "dude," and yet managed to have an intelligible conversation (for themselves and the listener).

4. Neandertal speech regularly used questions, commands, exclamations, and perhaps directional reference (indicatives). The differences may have been marked with "aspect" words, or morphological rules, or even grammatical rules. But the difference might also have been delivered through context or change in tone of voice, or even gesture.

5. Neandertal speech was capable of describing new situations, as when they juxtaposed terms for animals and places that they had not combined before. So at least in this sense, their speech must have been able to express new thoughts, what the linguists call productivity. But given its likely heavy reliance on long-term memory, their productivity was probably limited to reshuffling a very large body of lexical elements and phrases.

This communication system would have been capable of delivering a very large amount of information in context, but would have been less capable than modern language of long-range reference in the absence of appropriate cues. Neandertal language was direct and task-relevant. It was capable of referring to events in the past, or future, or at distant places, but only in ways connected to a context shared with the listener. There is no reason to think that Neandertals created elaborate stories or myths. Recall that Neandertals appear not to have used fire in the same social way that modern humans do. Moreover, they had few interactions with neighboring territorial communities, and therefore no reason to have modes of speech that could be used to interact with strangers or even acquaintances.[12]

The picture of Neandertal speech we have just presented is a minimal one based on what we know about Neandertal life from the paleoanthropological record. Neandertal speech may have been more powerful and subtle than this picture suggests. It may have included features quite foreign to modern language that evolved in the Neandertal lineage since the time of *Homo heidelbergensis*. There just is no evidence that requires anything beyond the features we have presented.

Further Reading

1.   Cheney, D., and R. Seyfarth. 1990. *How Monkeys See the World: Inside the Mind of Another Species*. Chicago: University of Chicago Press.

2. Savage-Rumbaugh, D., S. Shanker, and T. Taylor. 1998. *Apes, Language, and the Human Mind*. New York: Oxford University Press.

3. Noble, W., and I. Davidson. 1996. *Human Evolution, Language, and Mind: A Psychological and Archaeological Inquiry*. Cambridge, U.K.: Cambridge University Press.

4. Lieberman, P. 1984. *The Biology and Evolution of Language*. Cambridge, Mass.: Harvard University Press.

5. Lieberman, P. 1989. The origins and some aspects of human language and cognition. In *The Human Revolution: Behavioural and Biological Perspectives in the Origins of Modern Humans,* eds. P. Mellars and C. B. Stringer. Edinburgh: Edinburgh University Press.

6. Holloway, R., D. Broadfield, and M. Yuan. 2004. *The Human Fossil Record*. Vol. 3, *Brain Endocasts: The Paleoneurological Evidence*. New York: Wiley-Liss.

7. d'Errico, F., et al. 2003. Archaeological evidence for the emergence of language, symbolism, and music: An alternative multidisciplinary perspective. *Journal of World Prehistory* 17(1): 1–70.

8. Danesi, M. 2008. *Of Cigarettes, High Heels, and Other Interesting Things*. New York: Palgrave Macmillan.

9. Everett, D. 2005. Cultural constraints on grammar and cognition in Pirahã. *Current Anthropology* 46:621–646. See www.icsi.berkeley.edu/~kay/Everett.CA.Piraha.pdf

10. Bickerton, D. 1981. *Roots of Language*. Ann Arbor, Mich.: Karoma.

11. Bickerton, D. 2008. *Bastard Tongues: A Trailblazing Linguist Finds Clues to Our Common Humanity in the World's Lowliest Languages*. New York: Hill and Wang.

12. Ambrose, S. 2001. Paleolithic technology and human evolution. *Science* 291:1748–1753.

7

A Neandertal Walked into a Bar...

*"It's not just that he walks upright and uses
complex tools. He also makes me laugh."*

COULD NEANDERTALS SMILE or laugh? Could they tell jokes? One major
difficulty in any speculation about Neandertal humor is that much modern
humor hinges on features of language (puns, plays on words, malapropisms,
etc.), and as we discussed in Chapter 6, Neandertal language almost certainly
differed from ours. But humor includes nonlinguistic varieties. Smiling and
laughing do not require language; human babies can smile at 2 months, and
they can laugh (or giggle) as young as 4 months to a good game of peekaboo
(which can make adults laugh too). What about chimpanzees and monkeys?
Can they smile and laugh? Yes, chimpanzees and monkeys have a distinct

smiley face,[1] which looks nothing like their angry face. Obviously they do not tell jokes, but they do make each other laugh by tickling, teasing, and playing.

So our first conclusion is that Neandertals very probably could smile and laugh, and it probably happened when they were tickling each other, teasing, or playing. In fact chimpanzees laugh when they are tickled in the same places as humans, so Neandertals probably laughed when tickled in their armpits and on their bellies and the bottoms of their feet. Can you imagine a Neandertal blowing on another Neandertal's tummy to make him or her laugh? And yet it probably worked.

But Could Neandertals Tell or Make Jokes?

The tougher questions are whether Neandertals could intentionally create a humorous situation that would make the observer laugh (acting funny but without tickling, teasing, or playing) and whether could they tell jokes. We might call the former humor *physical humor* or *slapstick*, which would more closely resemble the funny antics of the Three Stooges, who poked and hit each other and caused funny "accidents" like falling down and breaking furniture. There is virtually no scientific literature on the possibility that Neandertals used this kind of humor, with one exception. The archaeologist Steven Mithen proposed that Neandertals would have been incapable of appreciating certain kinds of humorous stories because their memories for different kinds of knowledge were isolated cognitively and could not be directly compared.[2] Let us explain.

> A kangaroo walks into a bar and orders a beer. The bartender gets the beer and then says, "That'll be ten dollars." The kangaroo pays but says nothing. Finally, curiosity gets the best of the bartender and he says, "We don't get many kangaroos around here." The kangaroo says, "At these prices, you're not going to get many more."

We have paraphrased a very old joke that the anthropologist Elliot Oring retold in his book about humor in 1992.[3] Oring was apparently the first scientist to use the phrase "appropriate incongruities." What he meant is that some parts of a joke are normal or appropriate, and other parts are weird, strange, or incongruous (inappropriate or not consistent). When these parts are joined together, it creates an appropriate incongruity, which, interestingly, humans often *naturally* find funny. Some scientists who study humor suggest that it is the surprise element that helps make the joke funny, at least initially. This surprise element may be related to "surprise" games, like peekaboo, that made us laugh as children. So when a person says, "A kangaroo walks into a bar…," listeners are initially surprised and begin to get into the mood for an even more surprising ending.

Why Would Humans Naturally Like Surprises?

It is interesting that humans appear to like surprises. It is even more interesting that we seem to enjoy being scared, at least when we know that we cannot actually be hurt, as when we watch a scary movie or a child responds to a jack-in-the-box. Psychologists who study humor have long noted the element of surprise in jokes, but they have not often speculated about why surprise is prized. The linguist Jean-Louis Dessales has noted that people tend to like other people who surprise them with a recent news event or a juicy bit of gossip or a joke.[4] He hypothesized that in the evolutionary past it was important *not* to be surprised, so we tended to surround ourselves with people who could alert us to danger. According to Desalles's theory, we now appreciate people who surprise us because we have always favored people who can keep us up to date and well informed. Thus it may be that we now like jokes that surprise us (and the people who tell them) because we unconsciously favor anyone who is able to anticipate danger.

Returning to the kangaroo joke, there is a surprise right in the beginning ("A kangaroo walks into a bar") and a surprise at the end ("At these prices, you're not"). Of course, the surprise is maintained and even enhanced throughout the joke: we are surprised that a kangaroo can walk into a bar because we think they hop to get anywhere; we are surprised that the kangaroo can speak English and order a beer; we are surprised that the kangaroo knows about money and can pay for a beer. But the biggest surprise is not that a kangaroo can walk, talk, and order a beer, but that it knows $10 is too much for a beer!

Mithen suggested that Neandertals had a mind that was organized differently than ours, but he did not describe this difference by using terms from modern linguistics (language) or brain anatomy. So he really was not very specific about why Neandertals' knowledge of kangaroos would stay isolated from their knowledge of other things. Let us see if we can be more specific.

Modern Studies of Humor

Psychologists have studied humor for a long time, often tracing their interest in humor to Sigmund Freud. Freud collected jokes for years but had not done anything special with them until a close friend suggested that many dreams (which Freud had published a book about) seemed full of jokes. Apparently Freud had missed this connection, and in 1905 he published a book of his thoughts about humor entitled *Jokes and Their Relation to the Unconscious.*[5] He categorized jokes into two categories: mean jokes (he called them tendentious) and innocent (nontendentious). Freud appeared to be more interested in the mean category, and he thought that people told jokes in this category in order to trick (dupe) the listener or to express unconscious anger toward the listener. He thought that the joke teller was belittling the listener and making him or her feel "small," stupid, inferior, or tricked, all the while making the joke teller "superior." And, of course, Freud thought dirty jokes expressed the joke teller's sexual urges, either to humiliate sexually (to embarrass) the listener or simply to expose the listener to the joke teller's sexual thoughts and ideas without directly touching the listener. Here's the earliest dirty joke one of us (FLC) can remember (maybe when I was 7 years old?):

"Do you want to hear a dirty joke?"
"Yeah!"
"A white horse fell in the mud."

Notice that the listener expects to share some forbidden sexual idea with another person. Yet the listener is tricked, and the joke is disappointingly not sexual at all. In fact the only humor in the joke is that the listener has been tricked by the joke teller. Freud noted that there is only one recourse for the fooled listener: tell the joke to someone else. By tricking someone else with the joke, the joke teller gains superiority over the new listener. That was another observation made by Freud about jokes over 105 years ago: they spread quickly, "passed from one person to another like the news of the latest victory."

Theory of Mind

At the beginning of this discussion, we intentionally distinguished between telling jokes and making jokes. We suggested that making a joke might involve physical humor, like intentionally falling down or pretending to drop something valuable, and we suggested that Neandertals might laugh when Ogg (let us assume Neandertals had different names for each other) accidentally stepped into a fire. For one thing, as psychologists and Freud noted, we sometimes laugh at things that make us feel superior to someone else. But would

Ogg have been able to pretend intentionally to step on a hot coal and then dance around to the laughter of other Neandertals? To do so, he would have to have a thinking capability related to language but not language itself, that is, *Theory of Mind*.

As we have discussed throughout this book, Theory of Mind is the curiously named ability to guess what someone else is thinking. Some scientists call it "mind reading," but that has unwanted associations. Theory of Mind is a capability that most humans have and are fairly successful at. People with normal Theory of Mind are able to guess correctly that if they say something a certain way, the other person will react predictably. Often, when people think about how to say something (inner speech), they also anticipate what the listener's reaction will be. Thus normal people say things that do not often offend other people, scare other people, or make other people angry because they have thought through the various possible responses of the listener to the various ways of saying something. Normal people typically choose a way to say something that makes themselves appear normal, happy, and nonthreatening. Of course, normal people sometimes must defend themselves, and then they still sort in their minds among various ways of saying things to get a particular reaction. Perhaps the speaker is angry at the listener, but not too angry, so the speaker will sort through various ways of expressing anger and choose a particular phrase to avoid making things worse. This particular type of speech is also called "diplomatic speech" or "indirect speech."[6,7]

Does Joke Telling Require Theory of Mind?

Adequate or normal Theory of Mind is probably not the only mental requirement that joke telling requires, but it is probably a very important one. It has been noted that some children and adults with autism (a severe psychological disorder that disrupts thinking and emotions) or schizophrenia (a thinking disorder with delusions and hallucinations) have a very difficult time guessing what other people are thinking.[8,9] It is as if they are so restricted to their own worlds that they cannot possibly imagine what someone else is thinking. Thus people with autism and schizophrenia sometimes make others laugh at what they say, but sadly they often do so unintentionally. When they do tell jokes, they sometimes misjudge the reactions, and so they may offend others by saying something inappropriate (e.g., sexual) or scary. It has also been documented that a majority of individuals with autism or schizophrenia laugh more often for apparently no reason (or no reason another person can understand). However, some people with autism or schizophrenia who have high IQs (as in Asperger's disorder or high-functioning autism) lack Theory of Mind or are very poor at it, although they can appreciate normal humor, including sophisticated puns. But we are not suggesting that Neandertals were autistic! We're just speculating that joke telling may have required more sophisticated or higher level Theory of Mind than Neandertals were capable of, just as many contemporary humans who lack full Theory of Mind often are impaired in their ability to appreciate and create many types of humor.

What about Neandertals' Theory of Mind?

As we noted in Chapter 6, we think that Neandertals had language, but we think it may have been different from ours. We suggested that it might have contained a large number of words for places, routes, techniques for making things, other people, and emotions (then again, maybe not a lot of words for emotions, if they were as stoic as we speculated and their lives were very hard, as many scientists suspect), which were largely based on long-term memory. If this is true, then simple words identifying places, routes, techniques, and people would not easily lend themselves to joke telling. Of course, they still might have been able to make puns on these words, for example, if the name of a bad or stinky place (or if such a name were mispronounced) was also the name of a person they knew. With regard to emotions, it appears that Neandertals had emotional lives (as we speculate in a later chapter); however, it is the depth and breadth of their emotional lives that may be called into question. Is it possible they were stoic and had fewer words for emotions than *Homo sapiens sapiens,* yet had the same depth and breadth of emotions? Perhaps Neandertals were just not as touchy-feely as *Homo sapiens sapiens*. Or not. We shall discuss the possibilities of their emotional lives later.

We also suspected that their language might have had more than an average number of stock phrases (e.g., "If it ain't broke, don't fix it"). Part of our speculation here has to do with their generally low levels of innovation and an emphasis on repetitive expert procedures. As there is a consensus among scientists that this hypothesis is more true than false, we think that Neandertals' language might reflect the same lack of innovation. This lack of innovation coupled with stock phrases would probably not be helpful for making or telling jokes. It is also possible that storytelling and joke telling might have been the role of an older generation, and given that older Neandertal folk might have been in short supply, funny stories might also have been in short supply.

We have also suggested that Neandertals did not appear to use fire in the way modern humans did: sitting around the fire in a group, telling stories, sharing creation myths, and so on. They did not readily interact with neigh-boring communities of Neandertals or other human types. And they did not appear to trade with other communities. All of this evidence suggests that they may not have had the motivation to do so, either being afraid of strang-ers, lacking the curiosity, having no strong social need to interact, or maybe just being unable to convey their needs and wants to others. Let us take each of these in turn.

It seems unlikely they would be physically afraid of *Homo sapiens sapiens* because Neandertals were probably much stronger. However, recall the Neandertal from Shanidar who had a spear tip stuck in his ribs.[10] The nature of the wound suggests it was a *Homo sapiens sapiens* who threw the spear, as Neandertals did not have such long-distance weapons. The devices that propelled spears farther and faster were called *atlatls* (rhymes with *cattle-cattle*). So maybe Neandertals did fear *Homo sapiens sapiens,* not because of the way they looked, but because of the way they could throw spears long distances.

It is also possible that Neandertals did not interact or trade with others because they lacked curiosity. Here there is very little evidence to go on. In our recent unpublished survey that explores the common perceptions and biases of 165 undergraduate college students about Neandertals (originally conducted to see if anthropology majors' biases about Neandertals differed from psychol-ogy majors'; they did not), 74 percent of the sample thought *Homo sapiens sapiens* was more curious than Neandertals, while only 7 percent of the sample thought Neandertals were more curious than *Homo sapiens sapiens.*[11] Also, 82 percent of the sample thought *Homo sapiens sapiens* were more talkative than Neandertals, while only 2 percent of the sample thought Neandertals were more talkative. But what happens if Neandertals were simply unable to make themselves understood by others? This might occur if the language of a par-ticular Neandertal community, including their facial and hand gestures, was unique to that group. A recent study of deaf children raised together may help to support this argument.[12] These deaf children, who grew up isolated from anyone who knew a sign language (e.g., American Sign Language, British Sign Language, French Sign Language), together developed their own unique sign language and gestures. The instinct for language (or the instinct to communi-cate with others) is very strong in contemporary humans, and it has probably been a strong instinct for hundreds of thousands of years. Thus it is possible that these small, isolated Neandertal groups developed their own language dialects, which would have prevented easy transactions and interactions even with other Neandertal groups. Some scientists have proposed the possibility of distinctive Neandertal dialects,[13] but only between *Homo sapiens sapiens* and Neandertals, not specifically among Neandertal groups.

Were There Neandertal Clowns?

Anthropological studies of traditional societies, even those without any written language, have revealed, without exception, the presence of laughing, humor, and joking. Even though there are probably a plethora of evolutionary reasons for laughing and humor, one likely reason is to reduce the potential for conflict and aggression among community members. Many studies of these traditional societies describe the presence of a tribal clown. Usually the clown's job is to act as a buffer between two people who are arguing, deflecting or reducing the anger and tension by getting the two to laugh or getting observers to laugh. This hypothesis seems reasonable because normal humans are sometimes prone to laugh when uncomfortable, scared, feeling guilty, or to save face. Even if there weren't Neandertal clowns, perhaps Neandertals could laugh at the same things traditional contemporary societies laugh at, like the behavior of children, mocking other people who are too serious, and mocking people who are in positions of dominance (behind their backs, of course).

Other Cognitive Requirements for Joke Telling

Besides Theory of Mind , what other brain requirements are there for telling jokes? Many humor theorists have suggested that it takes many of the more sophisticated aspects of modern human language and brain functions to produce a joke. For example, producing a joke with an appropriate incongruity ("a kangaroo walking into a bar") requires the mental appreciation that the two things do not go together (kangaroos and walking into bars). It has been suggested that this operation takes an exceptional amount of brain power, as one must instantaneously contrast what one knows about the first thing (kangaroos) with what one knows about the second (what can walk into bars). Furthermore one must know that the incongruity creates an impossibility, and because it is so silly, it is funny. Interestingly, though, not every object walking into a bar is automatically funny. "The planet Jupiter walks into a bar" is not as funny as "A kangaroo walks into a bar." So the person is telling a joke cannot just set up arbitrary incongruities to make people laugh. "A parrot walks into a bar" is almost as funny as a kangaroo walking into a bar. "The United States Constitution walks into a bar" is not funny. Remember,

Mithen proposed that Neandertals could not contrast two types of knowledge in order to discover the funniness of an incongruity. He proposed that Neandertals were cognitively inflexible, but he did not explain how or why they were so. Earlier in this book we proposed that a limited working memory capacity might account for Neandertal's apparent cognitive inflexibility. Working memory capacity is the amount of ideas one can keep in mind and manipulate at the same time and stay on task, in spite of inside and outside interference.[14] We think *Homo sapiens sapiens* had greater working memory capacity than Neandertals.

Some complex sentences produced by language rely on *recursion*.[15] At its simplest, recursion is embedding a phrase within another phrase. For example, "The boy whose dog ran away went home" is recursive because "whose dog ran away" occurs in the middle of the main sentence, "The boy went home." Some humor scientists believe that recursive thinking, if not recursive language, is an important part of verbal joke telling. In their view, long-distance dependencies are created between elements in the joke that must be kept together in mind before the full meaning and humor of the joke can be realized and appreciated. Thus "A kangaroo walks into a bar" immediately creates a dependency between the knowledge people have of kangaroos and what they know kangaroos are not capable of, namely, walking, talking, ordering a beer, and paying for a beer in a bar. This whole bizarre juxtaposition of many knowledge facts must be kept in mind (working memory capacity) and then contrasted against each other to see if it makes sense (it does not). Then our minds must make some decision (this is really crazy but funny), after which we decide, consciously or unconsciously, to laugh or not to laugh. So there are really two more brain functions operating in this regard. One is recursive thinking or recursive language, and the other is the executive functions of the frontal lobes. The latter functions are the attention-payers and decision makers of our minds. Interestingly, even though we might have found the joke funny, we might "decide" not to laugh because we do not wish to please the joke teller. Perhaps we have some prior grudge against him or her and we decide not to give the pleasure and reinforcement of our laughter. This suppression relies on inhibition, one of the most important executive functions. As we wrote earlier, humor really is a *complex* brain function!

How Human Infants Develop Humor

An old biological rule is "ontogeny recapitulates phylogeny"; that is, the growth and development of an individual organism (ontogeny) replays the evolutionary history of its species (phylogeny). So over 150 years ago, when scientists observed a fertilized human egg, they believed it first "looked" like a fish, then it appeared to become more mammal-like, and then it developed into something that looked more and more human. The theory has long fallen out of fashion, but scientists still think that ontogeny might some-times recapitulate phylogeny in some way. We invoke this possibility for the development of humor. Is it possible that the way human infants begin to develop humor reflects the way humor developed evolutionarily over millions of years? Certainly the way human infants develop humor is scien-tifically interesting and valuable regardless of whether or not it reflects the evolution of humor.

All normal human infants can smile by 2 months old, and by 6 months old they can smile reliably when seeing particular faces (mom, dad, etc.). As we noted earlier, human infants can also laugh in response to someone tickling their bellies and toes at about 2 months old, and laugh in games of peekaboo at about 6 to 8 months old. Beginning at about 8 months old babies can also get adults to laugh by repeating actions, like splashing or banging, and by 14 months old, 100 percent of parents report their babies can act like little clowns, eliciting laughs from adults. By 6 years old, children can begin to appreciate bizarre and incongruous jokes, although they cannot always explain what exactly is funny about them. By 7 or 8, their humor begins to resemble full adult humor.[16] If we combine this evidence with the evidence from chimpanzees' and monkeys' humor (the comparative method), we can gain some insight into Neandertal humor. Undoubtedly Neandertals could smile and laugh at early ages. They could also probably tease and tickle, but they probably confined it to Neandertals their own age, siblings, or those of the same social status. In other words, Neandertal children probably did not go about teasing or tickling the dominant Neandertal (if there was one). By 7 or 8, they were fully capable of producing or understanding most adult Neandertal jokes—if adult Neandertals told jokes.

But there might have been a designated Neandertal clown, someone who provided comic relief in the tense situations when Neandertal communities came together. Here is our reasoning. Humor has an inherited component that varies in human individuals. If Neandertals' brain functions permitted them humor appreciation (at the very least), then at least one Neandertal with a strong tendency to making others laugh (since babies can do it) might have become the Neandertal clown. This clowning could have been confined to physical humor, thus bypassing any complex brain function requirements (high-level Theory of Mind, recursive thinking, executive functions of the frontal lobes, etc.). That way, inappropriate acts, like putting a potty chair on one's head or whatever the equivalent would be in Neandertal life, might have made other Neandertals laugh without having to resort to verbal joke telling. This physical humor in a community might accomplish the same ends as sophisticated verbal humor, such as calming people down, reducing the likelihood of aggression, providing comic relief from a stressful life, laughing at someone seemingly more stupid than you believe you are, mocking dominant individuals, and so on. Of course, it is only fair to point out that there is no evidence of clowns or clowning among *Homo sapiens sapiens* living at the same time as Neandertals; we have only anecdotal evidence for clowns and clowning around in contemporary foragers.

Humor and Group Size

The anthropologist Robin Dunbar suggested that the development of language replaced physical grooming as a bonding technique and reduced interpersonal aggression when ancient hominins made the transition to life on the ground and group sizes increased dramatically compared to the size of groups in trees.[17] We've already noted that it appears Neandertals lived in much smaller face-to-face groups than *Homo sapiens sapiens*. If it is true that humor, as an aspect of language, also helped to enhance social bonds and reduced the possibility of interpersonal aggression, then it seems possible that humor might have been much more important in larger than in smaller groups and hence more important to *Homo sapiens sapiens* than Neandertals. First a larger group might mean a greater potential for violent interactions. Second, a joke made or told in front of a group might have greater benefits for individuals with a tendency for humor. Third, laughing at someone's jokes is a way of making friends and showing the joke teller that the listener is trustworthy.

SO WHAT CAN we conclude? We have reviewed a number of possibilities for Neandertals' ability to create and appreciate humor:

- Just like *Homo sapiens sapiens* infants and chimpanzees, Neandertals could smile and laugh when they were tickled, teased, or when playing with other Neandertals.
- However, they could not make jokes or tell jokes. They lacked either the motivation or the ability to make others laugh (which probably means, among other things, that they did not have a well-formulated Theory of Mind).
- They may have had clowns who would make others laugh.

Further Reading

1. Preuschoft, S. 1992. "Laughter" and "smile" in Barbary macaques (*Macaca sylvanus*). *Ethology* 91:220–236.

2. Mithen, S. 1996. *The Prehistory of the Mind.* London: Thames & Hudson.

3. Oring, E. 1992. *Jokes and Their Relations.* Lexington: University Press of Kentucky.

4. Desalles, J.-L. In press. Have you anything unexpected to say? The human propensity to communicate surprise and its role in the emergence of language. In *The Evolutionary Emergence of Human Language,* eds. R. Botha and M. Everaert. Oxford: Oxford University Press.

5. Freud, S. 1960. *Jokes and Their Relation to the Unconscious.* Trans. J. Strachey. New York: Norton. (Original work published 1905).

6. Pinker, S., M. A. Nowak, and J. J. Lee. 2008. The logic of indirect speech. *Proceedings of the National Academy of Sciences* 105:833–838.

7. Coolidge, F. L., and T. Wynn. In press. The cognitive prerequisites for a language of diplomacy. In *Handbook of Language Evolution,* eds. K. R. Gibson and M. Tallerman. Oxford: Oxford University Press.

8. Baron-Cohen, S. 1995. *Mind Blindness: An Essay on Autism and Theory of Mind.* Cambridge, Mass.: MIT Press.

9. Brune, M. 2005. "Theory of Mind" in schizophrenia: A review of the literature. *Schizophrenia Bulletin* 31:21–42.

10. Schmitt, D., S. E. Churchill, and W. L. Hylander. 2003. Experimental evidence concerning spear use in Neandertals and early modern humans. *Journal of Archaeological Science* 30:103–114.

11. Coolidge, F. L., and G. Tambone. 2009. Perceptions of Neandertals' personalities. Unpublished manuscript, Department of Psychology, University of Colorado, Colorado Springs.

12. Senghas, A., S. Kita, and A. Ozyurek. 2004. Children creating core properties of language: Evidence from an emerging sign language in Nicaragua. *Science* 305:1779–1782.

13. Lieberman, P. 2000. *Human Language and Our Reptilian Brain.* Cambridge, Mass.: Harvard University Press.

14. Wynn, T., and F. L. Coolidge. 2004. The expert Neandertal mind. *Journal of Human Evolution* 46:467–487.

15. Coolidge, F. L., K. A. Overmann, and T. Wynn. 2010. Recursion: What is it? Who has it? How did it evolve? *WIRE Cognitive Science* 1:1–8.

16. Gervais, M., and D. S. Wilson. 2005. The evolution and functions of laughter and humor: A synthetic approach. *Quarterly Review of Biology* 80:395–430.

17. Dunbar, R. I. M. 1996. *Grooming, Gossip, and the Evolution of Language.* London: Faber & Faber.

8

To Sleep, Perchance to Dream

IMAGINE WHAT IT must be like sleeping in a tree. The wind blows, it rains, and occasionally predators prowl (and sometimes howl) beneath you. Would you try to sit with your back against the trunk of the tree and your legs dangling on each side of a branch? This is exactly the situation our distant ancestors faced millions of years ago. In order to solve the mysteries of tree sleep, today's anthropologists use the comparative method; that is, they look at animals most closely related to us, and see how they sleep. As you may be aware, chimpanzees are most like us, as we shared a common ancestor about 6 to 8 million years ago. So how do modern chimpanzees sleep? Well, they sleep in trees and make nests.[1] Anthropologists also suspect that our best known distant ancestor, *Australopithecus afarensis* (Lucy), probably slept in a nest also. And like modern chimpanzees, she probably socialized in nests, played in nests, and mated in nests. Sleeping in nests in trees had certain inherent dangers, such as falling out, that had a direct impact on the nature of sleep. But by the time of *Homo erectus* 1.8 million years ago, our ancestors had almost certainly given up sleeping in trees. In doing so they replaced one set of dangers with another.

Stages of Sleep

Have you ever started to fall asleep (driving or in your bed) and suddenly jerked yourself awake? The onset of sleep in nearly all primates is often accompanied by this reflex. It is called the "hypnic jerk" (or "hypnagogic jerk"). It appears to be a sudden muscle reflex that frequently awakens the sleeper. No one knows why the hypnic jerk occurs, but one of us (FLC) proposed that it is an ancient reflex that may have had evolutionary, adaptive value (those that can adapt, survive) by having the sleeper readjust or review his or her sleeping position in a nest or on a branch in order to assure that a fall did not occur.[2] It may also have had adaptive value if the reflex occurred in response to the beginning of an actual fall. Perhaps it allowed the sleeper a chance to stop a fall or to grab on to something at the beginning of a fall.

During this initial period of falling asleep, humans and chimpanzees (and many other primates) exhibit a rhythmic brain wave called "alpha." Compared to other brain waves, alpha has a medium amplitude (the electrical strength of a brain wave) and medium frequency (8 to 12 Hertz, the number of wave peaks or cycles in a one-second period). No one can say for certain why alpha waves evolved or what they really mean, except scientists know that alpha waves most often accompany the relaxed time just before the onset of sleep, when a human or primate is relaxed, eyes closed but awake, and just about to slip into sleep. People are easily aroused from their alpha waves; in fact any concentrated mental effort (like figuring out 17 times 17 in your head) or opening your eyes seems to disrupt the symmetrical alpha waves. Another curious thing happens in the alpha state and in the time when alpha waves stop and a person starts sleeping: weird thoughts and ideas sometimes occur. This is called the "hypnagogic state," and we can remember these weird thoughts or ideas but only if we've been awakened right after them (like by a hypnic jerk). Nonetheless scientists believe that sleep has officially begun when the alpha waves diminish and the brain waves show evidence of theta waves (4 to 7 Hertz) and sometimes higher frequency waves (greater than 15 Hertz) with a lower amplitude than alpha. They label this sleep "N-1" (what used to be called Stage 1 sleep), which means it is the first non-rapid-eye-movement sleep. It is considered a light stage of sleep, mostly occurring at the beginning of a sleep period, though it also occurs throughout a sleep period. People are easily awakened from N-1 sleep, particularly if someone says their name or if their child cries out. When awakened from N-1, people frequently report thoughts and ideas, but not necessarily what they would call vivid dreams with lots of action and emotions.

After 10 or so minutes of N-1 sleep, people usually drift into N-2 sleep (formerly Stage 2), also considered a light stage of sleep (where we are easily awakened), and again people report thoughts and ideas when awakened. It is also easily identified on an electroencephalograph (EEG, a machine that measures brain waves) because of two other brain wave features: spindles and K-complexes. N-2 brain waves look just like N-1 brain waves (higher frequency, lower amplitude), except that about every 30 seconds, a sudden, brief high-amplitude 13-to-16-Hertz wave occurs. These are sleep spindles. They appear to originate below the level of the cortex from lower brain structures like the upper part of the spinal cord where it joins the brain (the brain stem). Again, no one knows why they occur, just that they are characteristic of the lighter stages of sleep. N-2 sleep stays fairly constant across the human life span, and it occupies about 50 percent of one's total sleep. It also occurs in almost all primates, including chimpanzees. The second brain wave feature of N-2 sleep is the K-complex, which is a solitary, very large amplitude negative deflection in the brain wave pattern. Again no one knows why K-complexes occur, but they also appear to originate in lower brain structures.

After 15 to 45 minutes of sleep, people enter the deeper or deepest states of sleep, which are easily recognized on an EEG. These high-amplitude and very slow waves (1 to 3 Hz) are called "delta waves" and are the chief characteristic of N-3 sleep (formerly Stages 3 and 4 sleep). N-3 sleep is also called "slow-wave sleep." Slow-wave sleep has been observed in virtually all animals, including mammals and reptiles. Humans are very difficult to wake up when they are in slow-wave sleep, and they nearly always deny having any dreams, thoughts, or ideas in this state. All primates have slow-wave sleep. Interestingly slow-wave sleep is the one sleep stage that changes as people age. Beginning at about the age of 60, slow-wave sleep begins to diminish, at least in terms of the amplitude of the waves. There seems to be no mental decline associated with a lack of slow-wave sleep, and it appears that older persons simply spend more time in lighter stages of sleep, like N-1 and N-2. They also appear to stay in bed as long as younger people (on the average), but they seem to have more brief awakenings during their sleep.

After about an hour and a half, the EEG shows a curious pattern: the brain waves begin to look like N-1 sleep again (theta waves mixed with some higher frequency but lower amplitude waves), and until 1953 that is what sleep scientists thought they were. However, when two sleep researchers at the University of Chicago pasted electrodes to the skin above and the skin outside of infants' eyes, they found that the eyes appeared to be rapidly moving back and forth under the closed eyelids. This phenomenon became known as "rapid-eye-movement sleep" or "REM sleep." When the same studies were conducted with adults, scientists found that this was the stage when vivid dreams occurred, along with strong emotions, physical signs of sexual arousal (but not necessarily psychological sexual arousal), and rapidly changing heart rate and blood pressure. REM sleep has been observed in all primates and virtually all mammals. Interestingly the platypus has the greatest percentage of REM sleep of the total sleep of all mammals, about 33 percent. In monkeys REM sleep takes up about 5 to 15 percent of their total sleep. In chimpanzees it takes up nearly 20 percent of their total sleep (at least for captive chimpanzees sleeping in cages). In humans it takes up about 25 percent of sleep throughout the life span, although it occupies about 50 percent of the total sleep of newborn infants and babies.

What Is So Dangerous about Slow-wave Sleep?

Remember, the simple onset of sleep is accompanied by muscle relaxation, which is already a dangerous proposition for anyone planning to sleep high in a tree to avoid predators. The onset of slow-wave sleep is dangerous because it is such a deep state of sleep; we are completely unaware of our external environment and are very difficult to wake up. Nonetheless nearly all animals have slow-wave sleep, so despite its obvious dangers, animals have figured out how to survive it. Is it possible that all older animals, especially older primates, have less slow-wave sleep or may have lighter slow-wave sleep so that they can awaken the younger animals who have longer periods of slow-wave sleep? Perhaps a face-to-face group is more likely to survive because the lighter sleep of its older members ends up protecting younger members.

What Is So Dangerous about REM Sleep?

REM sleep has its own unique dangers. It is a paradoxical sleep stage because it has some characteristics of deeper stages as well as some characteristics of lighter stages. One quirk of REM sleep is that sometimes the sleeper is aware of selective aspects of the environment, such as when one dreams of being in a cold cave when actually the furnace has shut off. This phenomenon has been called "REM incorporation," because the sleeper includes selected aspects of what is happening in the environment into his or her dreams.

An even more dangerous aspect of REM sleep is the muscle paralysis that occurs at the onset of REM. In order to prevent sleepers from acting out their dreams while sleeping, special brain cells (inhibitory neurons) begin to fire at the onset of REM. These cells stop the voluntary muscles from moving during REM sleep. (Heart muscles are involuntary, so they are not stopped from moving.) This muscle paralysis is called "muscle atonia," and it occurs in virtually all animals that have REM sleep. Now a sleeping organism is truly vulnerable in REM sleep, so one's only hope is to awaken or be awakened before a predator strikes because the sleeper is literally paralyzed! Perhaps some of you have experienced this phenomenon. You are trying to wake up, but you find you cannot move your muscles. Of course, typically you are not actually awake; you are simply dreaming that you are awake. REM incorporation occurs, and the muscle paralysis becomes part of your dream. One of us (FLC) has proposed this muscle paralysis as the origin for so many people's reports that UFOs have captured and paralyzed them while the aliens performed experiments on them.[2] First, these "abductions" are almost always reported as occurring during the night (when dreams occur). Second, most people who dream of UFOs are especially intrigued by them in their waking life (and Freud was at least partly right that we sometimes dream about things we wish to happen). Third, people swear the incidents are real, and they are, at least for them; when we are dreaming, we are usually totally involved in the reality of the dream. We rarely challenge a monster or predator or other evil thing in our dreams. When we are dreaming, we are usually completely caught up in the reality of the dream, so we do not fly off buildings to escape predators, run through walls (although we certainly sometimes do these weird things in dreams), and we do *not* constantly remind ourselves that "this is only a dream."

Given these dangerous aspects of sleep and the fact that humans cannot live without sleep, it must be that sleep is important and that probably these dramatic stages of sleep, slow-wave sleep and REM sleep, are also very important. Slow-wave sleep and REM sleep became safer when our most distant ancestors began sleeping on the ground instead of in a tree, because being totally unaware of the environment (in slow-wave sleep) or paralyzed (from REM sleep) would not have the bad consequences on the ground as it did in the trees. If this prediction is correct, then we might see an increase in either slow-wave sleep or REM sleep compared to our closest relatives. Indeed, this is exactly what is observed: humans have about 25 percent more REM sleep than chimpanzees, even though we have very similar lengths of sleep (around 8 hours), and humans have about double or triple the amount of REM sleep as monkeys. There is not as much evidence for slow-wave sleep differences between humans, chimpanzees, and monkeys, but preliminary findings seem to indicate less of a difference between the three species in terms of slow-wave sleep than for REM sleep. Modern humans appear to spend about 15 percent of their total sleep in slow-wave sleep, depending on their age and amount of exercise, as it has been found that exercise tends to increase the percentage of slow-wave sleep. Earlier in this book we established that Neandertals had greater expenditures of energy than modern humans because they had to move around stockier bodies. Is it possible that Neandertals had a greater percentage of slow-wave sleep than *Homo sapiens sapiens* because of this added exercise?

Was Neandertals' Sleep Similar to Ours?

Given that Neandertals and modern humans had a common ancestor much more recently than chimpanzees and modern humans, and given that our sleep is not grossly different from that of chimpanzees, there is no reason to think that Neandertals' sleep was very different from ours. However, because of their stockier body, they did have greater energy requirements than we do. When modern humans exercise more than they normally do, it does seem to increase the amount of their slow-wave sleep, at least for a night. So it is possible that Neandertals had more slow-wave sleep than we currently do. However, there is no reason to believe their REM sleep percentage was any different from ours (about 25 percent). What Neandertals dreamed about is a tougher question to answer, but let's first explore how Neandertals and ancient *Homo sapiens sapiens* might have benefited from sleeping on the ground.

What Do Better Sleep and More REM Sleep Have to Do with Anything?

We previously proposed that there may have been benefits from having longer, safer, less interrupted sleep periods by sleeping on the ground and sleeping with others compared to sleeping alone in trees. Two of these benefits come directly from the content of REM sleep itself (and perhaps the thoughts and ideas from N-1 and N-2 sleep stages too). The first of these benefits comes from the fact that about two-thirds of all dreams tend to be about bad or negative things. Why would that be? Over 100 years ago Sigmund Freud claimed that *all* dreams were wish-fulfilling. (He later changed his mind and claimed that *most* dreams are wish-fulfilling.) Do we actually wish bad things would happen? Probably not, so we must rely on a more current dream theory to explain why more dreams are negative than positive. A few years ago, the psychologist Antti Revonsuo proposed that dreams anticipate dangers we might face while we are awake. He called his idea the threat simulation theory.[3] He thought dreams simulated threatening events that might have occurred to us during our evolutionary past. For example, the young daughter of a colleague of ours came running out of her bedroom after a nap and said that a giant bee flew through the window and bit her on the arm. Her parents showed her that there were no bite marks on her arm and that the window in her room had a screen. Why, then, would she dream about being bitten by a giant bee? According to threat simulation theory, she was replaying the threats of insects, snakes, and other creepy-crawlies in the environments in which early hominins evolved. Other modern dream theorists, along with your authors, believe that some dreams not only may replay ancient threats to our survival but they may also help us to anticipate threats that might come from our present environment. As evidence of the former, many children's dreams, up to about 50 percent, contain animals. More interestingly these animals are ones not often encountered in the children's actual lives: bears, wolves, snakes, gorillas, tigers, lions, biting insects, and mixed-up creatures that are half-man, half-wolf, or 8-foot-tall robot rabbits with antennae (FLC's childhood nightmare).

But probably the strongest evidence that dreams replay ancient threats comes from the two most common dream themes in the world: falling and being chased. We frequently ask our students how likely is it that in their present lives they would fall to their death. Of course, the answer (even here in mountainous Colorado) is that it is not at all very likely. And yet people all over the world dream of falling. Why? To us, it is obvious: as primates we spent tens of millions of years living and sleeping in trees! It was a real threat for us for perhaps our past 25 to 60 million years as primates. So we can imagine that ancient hominins dreamed of falling. And how would that dream have helped them survive? Well, imagine that you live and sleep in trees, and one night you dream that you are falling. Wouldn't you be more careful the next day making your nest, securing your nest in the branches, and in general being a little more careful the next night before falling asleep? And if modern *Homo sapiens sapiens* dream of falling, would not also Neandertals have dreamed of falling?

The next most popular dream theme is being chased; it is one of the most common dreams of female college students. Now Freudians might propose that it is a wish-fulfilling dream and that female college students unconsciously want to be chased. But again, if we go with more current dream theories, we might propose that dreaming of this ancient threat (back when little Lucy was being chased by all sorts of predators bigger and more dangerous than she) might have helped Lucy be more prepared in her ancestral environment, just as having that dream now might help prepare a female college student for similar threats in her current environment, encouraging her to get an escort to walk her to her dorm at night or to jog only along safe paths. Even male college students dream of being chased or hunted. By having this ancient dream now, in our present environment, we are all better prepared against real threats of being chased or hurt by real villains (murderers, father-stabbers, mother-rapers, etc.). According to Revonsuo, the evolutionary advantage of this kind of dream would help us better to avoid threats when we are awake. Psychologists also call these threat-avoidance skills "priming" because repeated practice (having the dream a lot) gets us ready (primes us) to be even more ready should an actual threat occur in our waking lives.

In his classic book *The Interpretation of Dreams*, published in 1900, Freud presented two common dreams people still often have: the examination dream and the being-naked dream. In the examination dream, the dreamer becomes aware that he or she is not ready to take a test. Freud said that this dream may happen many years after one has left school. In the naked dream, the dreamer realizes that he or she has left the house without any clothes on. It would seem that, unlike the dream of being chased, these two dreams would require some language skills. The chase dream requires only pictures, of oneself and of a predator, but the dream of not being ready for a test is much more sophisticated, requiring a school, a test, and more. We think that both the examination and the naked dream could still have relatively ancient roots, perhaps tens of thousands or even hundreds of thousands of years old.

Let's ponder the roots of the examination dream first. You might argue that there had to be schools in order to have an examination dream. But let's imagine that the dream has a more ancient evolutionary basis, such as being prepared to go on a hunt. Imagine two dreamers about 160,000 years ago, getting ready to go the following day on a big hunting expedition that may last several days. One dreams that he (assuming that hunting with spears was more likely to have been performed by men) is already hunting and suddenly needs his spear. He reaches to grab the spear, and…whoops! It is not there! The hunter needs the spear now! The hunted beast suddenly rises up and begins to attack the hunter! The dreamer is awakened in a panic but realizes it was just a dream. Wouldn't you imagine that hunter is going to remember to take his spear when he leaves for the hunt the next morning? Now imagine the second dreamer, who peacefully dreams about pleasant nothings. Who is more likely to remember to take everything necessary to be a successful hunter? The hunter who had the anxiety-provoking dream about not being properly prepared, of course! Thus we see that the roots of the modern examination dream might have had a much longer evolutionary history than simply the start of schools and exams. And by having that dream, the dreamer is better prepared to be successful the next day.

We think the same is true of the naked dream. Again imagine two ancient dreamers sleeping before a hike or hunt the next day. One dreams she (not to imply that only women dream of being naked) is walking naked and it begins to snow. "Oh my, I have left the shelter without my deer skins! I will freeze to death," she thinks. The other dreamer dreams of pleasant nothings. Who will be more likely to remember to dress warmly the next morning, to remember everything she needs for a successful hike or hunt? We think these kinds of dreams prime the dreamer to be prepared in his or her subsequent waking life.

Creativity and Innovation

Another advantage of increased REM sleep in modern humans may come from the fact that dreams can be creative, unusual, or innovative.[4] It's interesting to observe that the link between creativity and dreaming is nearly as old as written records. The earliest writing dates back to almost 5,000 years ago.[5] And right from the beginning, they often deal with dreams. A Sumerian king in 2,200 BC preserved at least two of his dreams on clay tablets. They reveal that the king had a puzzling dream and that he searched for the meaning of his dream by asking for help from a goddess. This record serves as one of the first examples, in what was to be a rich history of creativity in dreams, that dreams may be divinely inspired and that answers to waking problems may be sought in dreams. Mesopotamians and ancient Egyptians had long traditions of dream incubation; patrons would go to an incubation temple to sleep and dream and subsequently have their dreams interpreted by an expert. Solutions to problems and even treatments for diseases were sought in dreams, and it was also believed that dreams could be used to prevent or change future misfortune.

There are all kinds of modern examples of the link between creativity and dreams. The artists Dürer, Goya, Blake, Rousseau, Dalí, and Magritte and the musicians Mozart, Wagner, and Billy Joel, among others, all found inspiration in dreams. The eighteenth-century violinist Tartini reported that his inspiration for his most famous violin work, "Trillo del Diavolo" (The Devil's Trill), came to him during a dream in which the devil played a particular violin riff. Upon awakening, he excitedly tried to duplicate the devil's trill. The Charlie Daniels Band has a song, "The Devil Went down to Georgia," about the same dream of a violin-playing devil. It has been long claimed that the musical riff to "[I Can't Get No] Satisfaction" came to the Rolling Stones' lead guitarist Keith Richards in a dream; recently he wrote that when he woke up one morning, he found that the riff had been recorded on his tape recorder, though he did not remember playing it.[6]

Countless writers too have claimed that the inspiration for a work came in a dream. Robert Louis Stevenson wrote that, while pondering a duality that exists in all humans, he dreamed the story for *Dr. Jekyll and Mr. Hyde* in virtually a single dream. In one of the most famous instances, Samuel Taylor Coleridge claimed that his poem "Kubla Khan" came to him in a dream, and that upon awakening he wrote down about forty lines before he was inter-rupted by someone, thus leaving the poem incomplete.

Two nineteenth-century chemists have anecdotally reported that their most famous discoveries resulted from dreams. The Russian chemist Dmitri Mendeleev said he conceived of the periodic table in a dream. Friedrich Kekulé reported being stymied by the fact that neither a branching chain nor a straight alignment would account for the chemical properties of the six-carbon structure of a benzene molecule. In a dream he saw snake-like "con-firmations" writhing together. He reported that one of the snakes had seized its own tail and "whirled mockingly" before his eyes. He awoke in a "flash of lightning" and began working out his famous solution to the problem, the benzene ring. Also in the 1800s, the American inventor Elias Howe reported that he had worked for five years trying to create an automatic sewing machine and yet was unable to figure out how to get the machine to grab the thread once it had pierced the material. In a dream, he was a missionary captured by natives who stood around him dancing with spears that had holes in their tips. Upon awakening, he recognized that this was the solution to his problem: a needle with a hole in the tip.

There are also many modern examples of creative ideas and solutions to problems arising from dreams. In a survey of contemporary mathemati-cians, over 50 percent reported that they had at least once solved a mathe-matical problem in a dream. The brilliant Indian mathematician Ramanujan (1887–1920) said that the goddess Kali gave him solutions to theorems in his dreams.

The American psychiatrist Ernest Hartmann believes that dreaming allows the dreamer to make connections between very different thoughts and ideas.[7] These connections, he proposed, are often more broad and wide-ranging than during wakefulness. Hartmann also believed that dreams are primarily guided by our emotions, and that by using established brain pathways for seeing and detecting shapes, the story of the dream becomes an explanation for the dreamer's emotions.

The preceding examples are all anecdotal and, though provocative, are not the outcome of controlled scientific investigations. Is there any current labora-tory evidence that sleep and dreaming help us solve problems? The answer is yes. In just such a study, 106 adults were given a task measuring how fast they could respond (reaction time).[8] If they learned a secret trick about the task, they could respond much faster. After initial training, they either slept for 8 hours or stayed awake. Twice as many adults who slept guessed the secret trick when they awoke, regardless of time of day, compared to those who stayed awake. Based on the faster reaction times in the sleep group, the researchers postulated that the participants' greater insight was not a strengthening of the procedural memory itself, but involved a novel *restructuring* of the original representations. Thus, in their opinion, sleep may serve as a catalyst for insightful thinking.

Sleep, Memory, and the Ability to Learn

Imagine learning to juggle three balls. It's a somewhat difficult task; only about one or two people out of a hundred can do it. Experimental psychologists consider it a *procedural* memory task or *motor* task (as opposed to learning the definition of a word, which they consider a *declarative* memory task). These two forms of memory are not just arbitrary. Learning a *procedure* and learning a *word* are very different, as we saw in Chapter 3. First, they are learned in different parts of the brain. Second, they are learned differently; most procedural tasks, like juggling, take lots and lots of practice, upwards of hundreds and hundreds of trials, while learning a new word might take only a single trial. Third, people can learn a procedural task by watching someone else do it. In fact giving people lots of verbal or written instruction doesn't really help them learn to juggle. However, you can simply read *"ek* means one in Hindi" and learn it without anyone showing you anything else; it doesn't take hundreds and hundreds of practice trials. But let's return to the example of learning to juggle.

The skill of juggling is probably best learned by *spaced* trials, that is, practicing for 15 minutes or so and then resting for a while (15 minutes, an hour, or a day), and then practicing again. Experimental psychologists believe that during the resting periods the procedural memory is being consolidated, that is, being processed and stored in the brain. Interestingly if a person were to undergo electroshock therapy right after either practicing juggling or learning what *ek* means in Hindi, the shock would interfere with the consolidation process, and the memory would be lost. In other words, the effect of practice would be gone. However, if after learning a procedure a person is not shocked, then the consolidation process can proceed normally. Experimental psychologists also believe that this consolidation process continues whether the person stays awake during the resting period or goes to sleep.

But here's the really interesting thing about sleeping right after learning a procedural memory: experimental psychologists believe that the procedural memory is not only consolidated but even enhanced by sleep, and they suspect it is slow-wave sleep (N-3) that helps the most in the procedural memory's enhancement.[9] In animals it appears that REM sleep may consolidate certain types of procedural memories, such as how to negotiate a maze.[10] And in humans there's evidence that slow-wave sleep and REM sleep help in learning maze-like computer tasks,[11,12] which may be related to humans' ability to learn the landscape of their environment. There's also some evidence that REM sleep may help one type of declarative memory, known as episodic memory or autobiographical memory. These are memories that are recalled like stories of important parts of our lives, with a specific place and time and often having emotions associated with their recall. Experimental psychologists are not all in agreement about whether it is specifically N-2, N-3, or REM sleep that helps to consolidate and enhance memories, and they also do not completely agree about what types of memories are consolidated and enhanced by sleep. However, there is some general agreement that sleep, and perhaps certain stages of sleep, do help to consolidate humans' memories. And because almost all humans tend to confine their sleep to a single long sleep period (about 8 hours), it may be that the sequence of stages and having a single, long sleep period aid in our ability to learn things.

Sleeping on the Ground Instead of in Trees Did What?

We think sleeping on the ground changed the nature of sleep. Clearly, from an evolutionary perspective, dream sleep seems to have increased. In addition, humans sleeping on the ground were less threatened by the environmental dangers of sleeping in nests high in trees, battered by high winds and driving rains. Furthermore sleeping on the ground lessened the dangers of deep slow-wave sleep and the muscle paralysis that accompanies REM sleep. There also may have been cognitive benefits for procedural memories of having slow-wave sleep followed by N-2 and REM sleep and of having a single long and uninterrupted sleep period. But perhaps most important, the vivid dreams in REM sleep of being chased or attacked may have helped early hominins to protect themselves from these threats upon awakening. Dreams of being naked or unprepared may have helped early hominins to be more appropriately dressed and more prepared before venturing out on a hunt. And finally, some creative thoughts and innovative ideas that come through dreams may have helped early hominins be more creative and inventive upon awakening. After all, can anyone say with certainty that creative thoughts and innovative ideas can *only* come to us when we are awake? As we already pointed out, the ideas for some songs, poems, mathematical ideas, the chemical periodic table, and the automatic sewing machine all came to people while they were dreaming.

What Did Neandertals Dream About?

Because *Homo sapiens sapiens* and Neandertals shared a common ancestor about 500,000 years ago, their sleep was most likely just like ours, with similar percentages of slow-wave sleep and REM sleep. Our common dream themes were probably theirs: Neandertals dreamed about being chased, being attacked, being naked (maybe) or unprepared (more likely), falling, animals, and other people. One difference may be that because Neandertals hunted more big game animals than did *Homo sapiens sapiens* they may have had more dreams of danger and threat. There is evidence from the dream reports of contemporary hunter-gatherers that daily dangerous confrontations do increase dream themes of aggression and anxiety. Dream studies conducted in the 1930s on the Yir Yoront, a group of Native Australian hunter-gatherers, found that the adult males had significantly higher percentages of dreams with animals, aggression involving animals, and physical aggression than did those of male American dreamers. Studies of the dreams of Mehinaku Indians of central Brazil revealed more aggression and animal aggression themes than those of American dreamers, and about 60 percent of the dreams of the Mehinaku males had threatening themes, while only 20 percent of their dreams involved nonthreatening or nonaggressive activities.[2]

But if Neandertals had similar percentages of REM sleep as modern *Homo sapiens sapiens,* why were they not as creative and innovative? This is a tougher question to answer. Perhaps Neandertals did have creative and innovative dreams, but upon awaking, their limited working memory capacity or limited use of language made them unable to translate these dreams into reality. Another possibility is more grim: Neandertals appear to have lived a rougher, tougher life than *Homo sapiens sapiens* living at the same time. It is possible that their dream content contained many more negative themes and had more aggressive content than *Homo sapiens sapiens.* So maybe Neandertals were either incapable of innovative and creative dreams, or their tough lives increased negative dreams and included fewer innovative and creative dreams. There is also a possibility that Neandertals had creative dreams but were unable to communicate their dreams and ideas to other Neandertals.

IN SUM, WE believe that by sleeping on the ground *Homo erectus* and their descendants lessened the dangers of deep, slow-wave sleep and the muscle paralysis that comes with REM sleep. Of course, this tree-to-ground life transition brought new dangers, but with these dangers came solutions that favored bigger brains.[13] It's clear that modern *Homo sapiens sapiens* has more REM sleep than our closest living relative, the chimpanzee, and much more REM sleep than monkeys (a more distant relative). And we already noted many recent scientific studies that have shown that slow-wave sleep, REM sleep, and perhaps other stages aided in the learning of procedural memories like stone-knapping and the ability to learn new territories very rapidly. Of course, Neandertals were expert stone tool makers,[14] so they too would have benefited from sleep stages that aided procedural learning. As we saw in Chapters 2 and 4, Neandertals moved about well-defined territories but rarely traveled outside of them, and *Homo sapiens sapiens* often did journey outside of their home territory to interact with neighboring groups. It is possible that *Homo sapiens sapiens'* wanderlust was abetted by a sleep profile that enhance their ability to memorize new lands routes and landscapes.

So what can we conclude about Neandertals' sleep and dreams?

- If it is true, as many kinds of evidence suggest, that REM sleep aids creativity and innovation, then perhaps Neandertals did not have as many creative and innovative dreams as *Homo sapiens sapiens*. Did limitations in their working memory capacity restrict their creativity in their dreaming in some way? It has been demonstrated that greater working memory capacity is related to more imaginative thinking and planning.[15]

- Or did Neandertals have just as many creative and innovative dreams as *Homo sapiens sapiens*, but upon waking could not translate these dreams into reality? The anthropologist David Lewis-Williams proposed that Neandertals could not apply the contents of their dreams to their daily life in practical ways.[16] Did something about their language restrict them from communicating their creative ideas to other Neandertals?

- Or perhaps Neandertals were *capable* of as many creative and innovative dreams as *Homo sapiens sapiens*, but their dreams had many more negative themes and had much more aggressive content than the dreams of *Homo sapiens sapiens*, which may have prevented the innovative and creative dreams from occurring. As you will see in the next chapter, there are examples of Neandertal cannibalism. Imagine your dream content if there was a chance you'd be eaten by a new group of Neandertals you had just encountered! Of course, humans very often dream about what they wish were true. Perhaps Neandertals had dreams of feasting! And starving people do dream about food more than well-fed people.[17] If Neandertals' diets were more dependent on big game than *Homo sapiens sapiens* were, then Neandertals might have gone hungry more often than *Homo sapiens sapiens* living at the same time. Food dreams may also have blocked the content of more pleasant and creative dreams.

So what do we believe? First, we firmly believe that Neandertals had less working memory capacity than *Homo sapiens sapiens*. As greater working memory capacity has been demonstrated to be related to novel problem solving, imagination, creativity, and planning, it seems very likely that these cognitive processes persist into sleep and dreams. Therefore, it seems likely that Neandertals did not have as many creative and innovative dreams as *Homo sapiens sapiens*. And the other factors we discussed may also be relevant. If Neandertals went hungry more often than *Homo sapiens sapiens*, then their dreams might have been dominated by food. If their lives were more harsh and difficult, their dreams might have had many more negative themes, which also might block any creative thoughts. Finally, it is also possible that upon waking, they might have had more difficulty communicating their dreams to other Neandertals. We speculated in an earlier chapter that Neandertal speech might have had a greater reliance on long-term memories; thus their language productivity might have been limited to reshuffling what they already knew. This possibility means that Neandertals might have indeed had trouble communicating brand new ideas and translating them into practical applications.

Further Reading

1. Fruth, B., and G. Hohmann. 1996. Nest building behavior in the great apes: The great leap forward? In *Great Ape Societies,* eds. W. C. McGrew, L. F. Marchant, and T. Nishida. Cambridge, U.K.: Cambridge University Press.
2. Coolidge, F. L. 2006. *Dream Interpretation as a Psychotherapeutic Technique.* Oxon, U.K.: Radcliffe.
3. Revonsuo, A. 2000. The reinterpretation of dreams: An evolutionary hypothesis of the function of dreaming. *Behavioral and Brain Sciences* 23:877–901.
4. Franklin, M. S., and M. J. Zyphur. 2005. The role of dreams in the evolution of the human mind. *Evolutionary Psychology* 3:59–78.
5. Van de Castle, R. L. 1994. *Our Dreaming Mind: A Sweeping Exploration of the Role That Dreams Have Played in Politics, Art, Religion, and Psychology, from Ancient Civilizations to the Present Day.* New York: Ballantine Books.
6. Richards, K., and J. Fox. 2010. *Life.* New York: Little, Brown.
7. Hartmann, E. 2000. *Dreams and Nightmares: The Origin and Meaning of Dreams.* New York: Basic Books.

8. Wagner, U., et al. 2004. Sleep inspires insight. *Nature* 427:352–355.

9. Walker, M. P. 2005. A refined model of sleep and the time course of memory formation. *Behavioral and Brain Sciences* 28:51–104.

10. Winson, J. 1990. The meaning of dreams. *Scientific American* 263:89–96.

11. Karni, A., et al. 1994. Dependence on REM sleep of overnight improvement of a perceptual skill. *Science* 265:679–682.

12. Peigneux, P., et al. 2004. Are spatial memories strengthened in the human hippocampus during slow wave sleep? *Neuron* 44:535–545.

13. Coolidge, F. L., and T. Wynn. 2006. The effects of the tree-to-ground sleep transition in the evolution of cognition in early *Homo*. *Before Farming: The Archaeology and Anthropology of Hunter-Gatherers* 4:1–18.

14. Wynn, T., and F. L. Coolidge. 2004. The expert Neandertal mind. *Journal of Human Evolution* 46:467–487.

15. Reuland, E. 2010. Imagination, planning, and working memory: The emergence of language. In *Working Memory: Beyond Language and Symbolism,* eds. T. Wynn and F. L. Coolidge. Chicago: University of Chicago Press.

16. Lewis-Williams, D. 2002. *The Mind in the Cave: Consciousness and the Origins of Art*. London: Thames & Hudson.

17. Brožek, J., et al. 1951. A quantitative study of perception and association in experimental semistarvation. *Journal of Personality* 19:245–264.

9

You've Got Personality

WHAT WERE NEANDERTALS like personally? If we use television commercials as a guide we know that they were just like us: they would like sports, motorcycles, and bowling. They would be attracted to the opposite sex, yet sometimes they would have relationship problems and get depressed. They would also be very sensitive and deeply hurt by any suggestions that they were a simple folk or that they were not very smart. Of course, television commercials are not a reliable guide. We can only speculate about what Neandertals were like personally from what we know about them from the paleoanthropological record, things like where they lived, where they traveled, what tools they made, what game they hunted and how they hunted, how they treated their dead, and other impressions they left behind.

Possible Neandertal Personality Trait #1: Pragmatism

In Chapter 5 we introduced the French archaeological site of Moula-Guercy, which had evidence for cannibalism.[1] The archaeological remains included two Neandertal adults (one very large and one smaller), two adolescents (about 15 or 16 years old), and two children (about 6 or 7 years old). Their skulls had been smashed, and there was evidence that their brains had been eaten. One of the children's tongues had been cut out. Their leg and arms bones had been stripped of their flesh, and the leg and arm bones had been broken or cut off. After the flesh had been cut off and the arm and leg bones had been cut up, the bones were then broken open for the soft marrow inside. Moula-Guercy presents the strongest evidence to date of the practice of cannibalism in Neandertals.

Now let's examine the psychological reasons for such behavior. First, cannibalism might be considered a means of getting food, known as gastronomic cannibalism.[2] Perhaps the cannibalistic Neandertals were starving, and these six Neandertals were the weakest, sickest, or most helpless. Remember, it appears that Neandertals lived in small face-to-face groups, and the cannibalized Neandertals at Moula-Guercy might well have been a single isolated family group. One of the six Neandertals in the Moula-Guercy family is among the largest Neandertals ever found, so unless he or she was sick, it seems unlikely this small group was completely helpless. They were probably overwhelmed or surprised by a group from a neighboring community, though we cannot eliminate the possibility that the diners were individuals from their own territorial group. The presence of all six cannibalized Neandertals suggests they were killed together, or were killed or died within a short time of each other, and that the purpose of their deaths was to get food.

But there is one important piece of evidence related to the gastronomic cannibalism hypothesis for Moula-Guercy: plenty of red deer were also butchered in the same cave, presumably at about the same time as the six Neandertals and in much the same way: their skulls were smashed, their flesh was stripped from their bones, and their bones were smashed to extract the marrow. Of course, finding other kinds of animal bones in addition to the bones of Neandertals doesn't eliminate the gastronomic cannibalism hypothesis; in fact it suggests that the cannibals were perfectly willing to butcher other Neandertals using normal butchery techniques. But it does seriously weaken the alternative hypothesis, that the cannibalism at Moula-Guercy was a form of funerary ritual.

Modern cannibalism mostly comes in two forms: survival cannibalism and ritual cannibalism. Here in Colorado we have one of the most famous cases of survival cannibalism from back in the days of the Wild West. The prospector Alfred G. "Alferd" Packer was accused of eating members of his group while lost and starving during a Rocky Mountain winter in the 1870s. (One of the dining halls at the University of Colorado–Boulder is named after Packer, and in 1993 *South Park* creators Trey Parker and Matt Stone created an independent film entitled *Cannibal, the Musical* based on Packer's story). But modern-day cannibals are more likely to be involved in aggressive ritual cannibalism, in which people eat their defeated enemies in order to gain some of their power or further degrade them. This seems unlikely for our Neandertal example, primarily because the corpses were treated just like the remains of prey animals, rather than in a special way that would suggest ritual. However, Neandertals' willingness to include fellow Neandertals in their diet plan does have implications for Neandertal personality.

Their decision about what to eat appears, at least at times, to have been driven by pragmatic considerations (the need for food) rather than sentimental ones (who got to eat and who got eaten). Cutting out a child's tongue to eat is the very opposite of being sentimental and appears to be good evidence for being practical, concentrating on the fact of needing to eat rather than the possible emotional consequences of cannibalism.

Possible Neandertal Personality Trait #2: Stoicism

Neandertals hunted big and dangerous animals; they ate meat and lots of it (over 80 percent of their diets) in order to fuel their large brains and stocky, heavy bodies.[3] It has long been established that Neandertal skeletons, even those of juveniles and sometimes children, show evidence of being injured at some time in their lives. And they often survived these injuries, as they show signs of healing. The skeletons of *Homo sapiens sapiens* living at the same time do not show as many injuries. Anthropologists have long thought that these differences suggest that life was more dangerous for Neandertals, and there was much greater risk of injury and more stress in their daily lives. In Chapter 1 we introduced the work of the anthropologists Thomas Berger and Erik Trinkaus, who compared Neandertal injuries to those of *Homo sapiens sapiens* living at the same time and of modern humans, including a sample of present-day North American rodeo cowboys. As they suspected, the pattern of Neandertal injuries best matched those of the rodeo cowboys: lots of head, neck, and shoulder injuries. If rodeo cowboys tend to injure themselves by getting up close and personal with large, aggressive animals, then Neandertals might also have spent a lot of time in very close physical contact when hunting large, dangerous animals. This suggests that Neandertals often had to endure pain, so it is very possible that this favored a stoic personality. Of course, most of their fellow Neandertals were occasionally also injured, so sympathy from others might have been in short supply.

Possible Neandertal Personality Traits #3: Bravery, Fearlessness, High Tolerance for Risk

Neandertals probably did most of their hunting with hafted spears or just wooden spears. To be effective these spears had to be thrust into the animals, not thrown at them. This would be consistent with their rodeo cowboy–like head, neck, and shoulder injuries. So what kind of personality traits would it take to be a Neandertal hunter? Certainly bravery would have to be at the top of the list. What about fearlessness? Just because a person is brave does not mean that he or she has no fears. So maybe Neandertals were brave but still anxious about going on a hunt, knowing full well that some of them might get hurt and some might be killed. Given their commitment to hunting the largest available animals (e.g., mammoths), it seems quite possible that they had evolved personalities that were less fearful than ours or assessed risk quite differently.

Possible Neandertal Personality Traits #4:
Charity and Sympathy

There is another important fact about injuries. If you recall, Shanidar #1 had survived very serious injuries that undoubtedly incapacitated him for an extended period of time. It appears that many Neandertals must have spent significant stretches of time convalescing. They must have had help.[4] Perhaps they were attended to by an individual who was experienced with treating wounds, broken bones, diseases, and other injuries. More likely, all adult Neandertals had some expertise in this domain given the frequency of injuries. Neandertals with such injuries must also have had some help from others getting food, water, and other necessities. Because such situations were common, it is very likely that Neandertal personalities included high degrees of empathy for other individuals.

Possible Neandertal Personality Traits #5:
Callousness, Coldness, Hardheartedness

There is another, more disturbing pattern of Neandertal injuries: there is little evidence that they survived serious leg injuries. Certainly Neandertals must have endured leg injuries in their very dangerous and up-close hunting strategies. So why are there so few examples of leg wound healing? It is possible that Neandertals placed high value on those members who could move about. Hunting and gathering were strenuous activities that often included walking many miles a day. Moreover close-in killing required agility as well as strength. A Neandertal with a head, neck, or shoulder injury could still move about and move about relatively quickly. But a Neandertal with a leg injury probably could not move about. Thus one possible explanation for the dearth of healed leg injuries among Neandertals is that these individuals were left behind or killed. This attests to Neandertal pragmatism, but also a certain degree of hardheartedness when situations required it.

Possible Neandertal Personality Traits #6:
Neophobia, Rigidity, Stubbornness, Traditionalism,
Dogmatism, Conservatism

In Chapter 1 we introduced another Shanidar Neandertal, an adult male who had a curious wound to the ninth rib on his left side.[5] When examined carefully, it appeared that his rib had been pierced by a sharp stone tip that had been fastened to a lightweight spear that was thrown from some distance. This technology, a stone-tipped but relatively light wooden throwing spear, has so far been found associated only with *Homo sapiens sapiens* and not with Neandertals. Neandertal spears were heavier and were clearly designed to be thrust from short range. The spear that hit this Neandertal also seems to have come from above. The most likely scenario for that type of wound would be a spear thrown slightly upward, which goes up and comes down (the ballistic trajectory). Neandertals did not use such weapons.

Whatever the specific scenario, it appears that a *Homo sapiens sapiens* attacked a Neandertal, and that the Neandertal survived—for a while. He survived at least two weeks but probably not more than two months, as the bone did not have time to completely heal. This leaves us with many mysteries, among them: Were Neandertals being hunted? Did the *Homo sapiens sapiens* attack in self-defense? This incident at Shanidar Cave poses more questions than it answers. It is interesting that Neandertals in the Middle East continued for several thousand years after this incident but never appeared to adopt the light throwing spear technology of *Homo sapiens sapiens*. Remember, Neandertals were expert stone knappers,[6] but their stone tools and heavier thrusting spears had been made much the same way for well over 200,000 years. Neandertals were not often creative. They made a spear that worked well, and they didn't change the design over time, despite interactions with *Homo sapiens sapiens,* who did have the lighter spears.

What can be said about the Shanidar Cave incident? It is all very speculative, but the incident appears to foreshadow Neandertals' ultimate fate. Neandertals and *Homo sapiens sapiens* shared parts of western Europe for several thousand years without the Neandertals borrowing much, if anything, from the more sophisticated technical systems of *Homo sapiens sapiens*. It is not that Neandertals could not have made such things if they had wanted; they were adept at working stone and other materials. They just did not adopt the new and more deadly technology even though it meant less risk. Does this mean Neandertals feared new ways of doing things, which psychologists call "neophobia"? (The psychologist Carl Jung called this fear of new things "misoneism.") The symptoms of neophobia include retreating from unfamiliar people or situations.

Neandertals were master stone knappers and expert hunters, but their success was accomplished by doing the same thing over and over again. They were almost certainly very dogmatic in their approach to life and tolerated no experimentation. We imagine that this extended to social life as well. A Neandertal father, if he were still alive, was unlikely to have turned a blind eye to the innovative shenanigans of his preteen offspring. He would have been a stern taskmaster.

Possible Neandertal Personality Traits #7: Lack of Autonoetic Thought

In Chapter 5 we reviewed the controversy concerning Neandertal mortuary treatments. Traditional interpretations of the Neandertals at La Ferrassie and Kebara Cave have been that they buried their dead. Close examination of the actual evidence, however, suggests that Neandertals' mortuary treatments were actually quite minimal. It does appear that they often took steps to protect a corpse, but the labor investment was usually low: covering the body with a few large rocks, or scooping out a very shallow depression in which to lay the body. This was not burial as we usually think of it and does not appear to have been based on an intention to protect corpses for long periods of time. Indeed Neandertal skeletal remains in many sites such as Kebara Cave and Krapina suggest that Neandertals often pushed aside skeletal remains. Use of corpse protection does suggest that Neandertals had emotional reactions to death and perhaps sentimentality toward the departed, but their interest and investment contrast strongly with the mortuary ceremonies of almost all humans living in the modern world, and even examples of *Homo sapiens sapiens* who lived in in Europe 27,000 years.

At the site of Sungir in Russia, archaeologists unearthed a series of burials that date back about 27,000 years.[7] The main Sungir burial consisted of an adult *Homo sapiens sapiens* and two youngsters, a boy about 13 and a girl about 8. The children were buried head-to-head. The boy's body was adorned with about three thousand ivory beads that were either necklaces or sewn onto his clothing. His grave also included a beaded cap. He was buried with a painted, stone pendant that probably hung from a string or leather rope around his neck. He also had painted ivory mammoth arm bracelets, a beaded cap with fox teeth, and a belt with about two hundred and fifty fox teeth. He wore an ivory pendant in the shape of an animal, an ivory sculpture of a mammoth was on his shoulder, and he had ivory ornamental disks. He was buried with an ivory hunting spear, which might have been tapered to the length of his body. The girl had a similar but smaller ivory spear and was buried with more than five thousand ivory beads.

The Sungir burials appear vastly different from the La Ferrassie and Kebara Neandertal burials. But what strikes us as the most interesting features of the Sungir burial are the ivory hunting spears. Because ivory is soft, such spears were unlikely to have ever been intended for real hunting; thus these were truly ceremonial spears. This implies that these *Homo sapiens sapiens* were thinking about a life after death, and perhaps hunting in the afterlife. Psychologists call this recognition that the afterlife may be quite different from regular life "autonoetic thinking."[8] When modern humans imagine that heaven is full of clouds, angels, and all (or some) of our dearly departed loves ones, they are engaging in autonoetic thinking.

Neandertals did nothing comparable. We think the apparent lack of auto-noetic thinking is a higher-order personality or cognitive trait that might have expressed itself in Neandertals' daily behavior, particularly their lack of curiosity about the past and the distant future (e.g., "Where do we go after we die?"), grounding their life in the here-and-now. Not worrying about the future may have made Neandertals less anxious or less depressed about accidents or mistakes.

Possible Neandertal Personality Traits #8: Lack of Diplomacy and Inhibition, Laconism, Xenophobia

As we saw in Chapter 4, Neandertals lived in small face-to-face groups and rarely traveled outside of their home territories. *Homo sapiens sapiens,* on the other hand, often acquired artifacts and materials from distant sources, either by trading for them or by making long treks. This suggests to many anthropologists that *Homo sapiens sapiens* must have had the communication and social interaction skills to deal with strangers and acquaintances. Such diplomatic or polite speech has some interesting requirements. Psychologists have also called it indirect speech,[9] which means it hides some real meaning or feelings but still gets people what they want. For example, you may not like the person who has something you want; if you ask for it directly or with hostility, you may not get it, but if you hide your true feelings and pretend that you *do* like that person, you may just get want you want.

Now, let's examine what this previous example of diplomatic speech requires. One of the most important things is the ability to inhibit your real feelings of dislike. Most nasty, antisocial, abrasive people do *not* hide their nasty or abrasive feelings, or perhaps *cannot* hide them. Diplomatic speech frequently requires hiding one's true feelings, and this "hiding" ability is one of the executive functions of the frontal lobes. Neuropsychologists call this ability "inhibition"; it's one of the most important executive functions, in addition to planning tasks, making goals, and developing strategies to finish tasks and attain goals. Another executive function is the ability to develop alternative strategies when the first strategy fails. So is it possible that Neandertals lacked the ability to talk diplomatically?[10] Did Neandertals have trouble inhibiting their true feelings? It is easy to imagine that both Neandertals and *Homo sapiens sapiens* would be fearful or distrustful of groups of strangers. But were *Homo sapiens sapiens* better at inhibiting their fear and distrust? Were they better able to speak in diplomatic ways than Neandertals? Is this why *Homo sapiens sapiens* 40,000 years ago were able to trade and barter for distant goods? Xenophobia is the fear and distrust of strangers, and it is clearly observed not only in contemporary humans but also in their closest nonhuman primate relative, the chimpanzee. Perhaps Neandertals not only lacked the ability to be diplomatic to strangers, but their inability was coupled with a natural fear and distrust of strangers.

TO SUM UP, archaeological and fossil evidence supports the following personality profile of the average Neandertal:

- Pragmatic, including callous when necessary
- Stoic
- Tolerant of risk
- Sympathetic and empathetic
- Neophobic
- Unimaginative
- Dogmatic and inflexible
- Xenophobic
- Direct but laconic

We imagine that most readers are now saying to themselves, "I know someone just like that." We certainly do. But an individual with all of these personality characteristics is in fact a bit unusual among modern humans—within the modern range to be sure, but not the norm. It was probably the norm for Neandertals. Of course, Neandertals' personalities varied, just as modern human personalities do. Personality is also susceptible to evolutionary forces, and Neandertal personalities evolved in circumstances that differed from those of our ancestors. Their personalities were not optimal for success at the United Nations or even at a large urban high school. But they were optimal for life in small groups in Ice Age Europe.

Were There Neandertals with Mental Disorders?

There is wide variation among modern humans on what it means to be mentally "normal" and have a "normal" personality. Much depends on one's age, gender, expected roles, culture, and subculture. However, one general rule of thumb holds true: you are normal if, after some stress or strain in your life, you do not tear off your clothes and go screaming naked down the street. In other words, you are normal if you can cope reasonably well with everyday stresses and strains. Notice that this definition also implies that everyday life is not perfect: it comes with all kinds of threats, stresses, and insecurities, but *most* people manage to get through it fairly well. The word "normal" is derived from ancient Latin (as is our word "norm"), and it means a rule or a standard. So in modern psychology, *normal* is what most people do in a given situation. To act differently is branded "abnormal" behavior, which has some dangerous implications, as acting differently from what most people would do in a situation could be noble, helpful, creative, or brave, such as running into a burning building to save another's life. Clinical psychologists have thus tended to get away from the terms "normal" and "abnormal" and use "disorder" instead, which denotes that the behavior is harmful to oneself or others and prevents the person with the disorder from keeping a job, staying in school, or having healthy relationships with other people.

The adjective "psychotic" describes someone who is out of contact with reality. In a very severe psychosis, a person may not be able to remember or say his or her own name. The most common type of psychosis is schizophrenia (accounting for over 75 percent of all patients in mental hospitals and about 1 percent of any population of people throughout the world). Schizophrenia has two main symptoms: delusions and hallucinations. Delusions are false beliefs, such as believing you are Jesus Christ or that aliens have been following you and are trying to take your brain. Hallucinations are perceptions without any real stimulation, such as seeing something that's not there, hearing voices when there are no voices, smelling or tasting something that's not really there (like a poison in the food), or feeling bugs or snakes crawling under your skin when there are no bugs or snakes under your skin. The most common form of hallucination is hearing voices, but interestingly, in schizophrenics who report hearing voices, the voices most commonly say everyday things such as "Brush your teeth" and "Get out of bed." Much less commonly (thank goodness!), the voices tell the person to kill someone. Schizophrenia is often accompanied by extreme religiosity in all cultures that have religions. When one of us (FLC) worked as a forensic psychologist in a maximum security mental hospital, about 80 percent of his patients were diagnosed as schizophrenic, and about 10 percent claimed to be Jesus Christ. When he worked as a psychologist in a southern India mental hospital, he had Hindu schizophrenic patients who claimed to be various Hindu gods. Interestingly, however, Muslim schizophrenics did not claim to be Mohammed.

The British psychiatrist Tim Crow argues strongly that schizophrenia, an officially recognized diagnosis, is restricted to beings with language.[11] Furthermore he proposes that schizophrenia is specifically created by problems in coordination between the left and right hemispheres of the brain. If true, it seems probable that Neandertals were prone to schizophrenia. As we saw in Chapter 6, Neandertals almost certainly had some form of language, and though it may have been quite different from ours, it was based in the same left hemisphere of the brain. This function evolved prior to the split between our ancestors and those of Neandertals. Thus just like anatomically modern humans 100,000 years ago, Neandertals were probably capable of becoming psychotic and capable of uncontrolled fits of rage (which, by the way are not classified as a type of psychosis in the current diagnostic system but as a separate disorder known as Intermittent Explosive Disorder). But Neandertals probably did not have delusions about aliens probing their brains at night or about being great gods; since they lacked innovation and creativity, they probably didn't believe in aliens or gods. There is also some speculation among psychologists that schizophrenic disorders are sometimes accompanied by creative ideas and thinking. Therefore, it seems possible that if Neandertals were less prone to schizophrenia, they might have also been less likely to take advantage of the creative thoughts and ideas of any of their brethren who were schizophrenic.

And what about neuroses? Were Neandertals capable of being anxious? Freud introduced the concept of neurosis and thought the chief symptom of a neurosis was anxiety, that is, worrying about something persistently and imagining a bad outcome. Psychologists nowadays still use the term, although not officially. A neurotic person is one who is unhappy, unsuccessful in relationships or work, yet without hallucinations or delusions (signs of psychosis). Neurotic people tend to distort and misinterpret reality. When a psychotic person walks by a group of people, he may think they are plotting to kill him.

When a neurotic walks by a group of people, she may think they are saying mean things about her.

Research into the behavior of nonhuman primates, especially those held in captivity, provides clear evidence that neurotic behavior is not limited to humans. Chimpanzees, for example, can exhibit strong symptoms of *some* kinds of neuroses. Of the official group of anxiety disorders, including phobias, panic attacks, and obsessive-compulsive disorders, all have been observed in caged and wild chimpanzees.[12] Therefore, we can safely assert that all hominins, including Neandertals, could have included neurotic members, who were overly fearful of strangers (social phobia), prone to sudden and overwhelming fear (panic attacks), or very attached to doing things the same way over and over and overly attached to routine ways of doing things (obsessive-compulsive disorder). Interestingly, another neurotic-like disorder, trichotillomania (pulling out one's hair to relieve stress or for pleasure), does occur in caged chimps. It is not clear whether it has been observed in wild chimps, but it's uncommon at the very least. It's another intriguing possibility for Neandertals (and *Homo sapiens sapiens* 100,000 years ago) if trichotillomania does occur in wild chimps.

Another major category of mental disorders is personality disorder. This grouping has been recognized for well over 2,000 years, as the ancient Greeks described many of the modern personality disorders around 350 BC. People with a personality disorder (they can even have two at once) mostly annoy or irritate other people. (But some are very dangerous to themselves and others.) People with personality disorders tend not to be bothered by their own behavior, and mostly they think they are just fine. They are not out of contact with reality, they do not *typically* have hallucinations or delusions (so they are not psychotic), and they are not *typically* very anxious (so they are not considered neurotic). However, it is clear that personality disorders are shaped by strong genetic factors, and they are very difficult to treat with psychotherapy. For one thing, people with a personality disorder usually don't seek therapy because they don't think they have a problem. If they are forced or cajoled into therapy by family or friends, they tend not to do well because the disorder is so much a part of their core personality.

One of the most dangerous personality disorders is the antisocial personality disorder. It used to be called "psychopathic personality" (and the person a "psychopath") and also "sociopathic personality" (and the person a "sociopath"). These people (more men than women get diagnosed) appear to have no conscience; tend to make terrible parents, neglecting or more often just abandoning their children; and often become con artists and take advantage of the elderly or people who are handicapped. They have great difficulty holding down steady jobs, and when they do, they often steal or embezzle from their employers, regardless of whether they work for a Fortune 500 company or a church, Goodwill, the Salvation Army, or the Muscular Dystrophy Association. Women with antisocial personality disorder usually don't murder other people, but sometimes they do, for example, poisoning their husband or boyfriend for money or drugs and sometimes for apparently no reason at all. Such people thus do commit antisocial acts: they murder, rape, steal, embezzle, abandon and neglect their children, take illegal drugs, and create mayhem. However, very often they manage to commit these acts by being *initially* charming and superficially responsible, warm, and loving. Then, just when others begin to trust them, they commit their antisocial acts. Some antisocial people are not charming and warm; they simply steal, rape, or kill when it's convenient for them or when it strikes their fancy. Evolutionary psychologists often label them "cheaters," as they rarely work long and hard for what they desire. They simply take it when they are ready.

Evolutionary psychologists have long recognized that many modern human thoughts and emotions are rooted in conflicts around cheating—not just the cheating of spouses, but also cheating by members of a social group who do not pull their own weight when it comes to hunting, gathering, foraging, helping, and being a good group member.[13] Since most anthropologists believe Neandertals lived in smaller face-to-face groups than *Homo sapiens sapiens*, it is likely that cheating was not as serious a social problem for them as it is was for *Homo sapiens sapiens*. Everyone in a Neandertal territorial community knew everyone else very, very well, so any individual prone to cheating would be found out very early and known as a cheater by everyone. Thus classic antisocial personality disorders may have occasionally occurred among Neandertals, but they likely would have been dealt with harshly early in life. *Homo sapiens sapiens*, with their large groups and distant social contacts, provided much more opportunity for an antisocial person's success, especially if he or she were of the charming variety.

There are many more official personality disorders, such as narcissistic personality disorder, in which people are power-hungry and highly manipulative; histrionic personality disorder, in which they are attention-seeking and sexually provocative; and schizoid personality disorder, in which they are extreme loners and have no need for other humans, even their own family members.[14] But we will not review all of them here. Suffice it to say that there is no reason to suspect that Neandertals could not have had some or all of these disorders, just like *Homo sapiens sapiens*. Scientists are currently completing the Neandertal genome. If and when scientists manage to identify the specific genome sequence for personality disorders in modern humans, all they will have to do is see whether these sequences are evident in Neandertal genes as well.

Further Reading

1. Defleur, A., et al. 1999. Neanderthal cannibalism at Moula-Guercy, Ardeche, France. *Science* 286(5437): 128–131.

2. Bocherens, H., et al. 2005. Isotopic evidence for diet and subsistence pattern of the Saint-Césaire I Neanderthal: Review and use of a multi-source mixing model. *Journal of Human Evolution* 49:71–87.

3. Spikins, P. A., H. E. Rutherford, and A. P. Needham. 2010. From homininity to humanity: Compassion from the earliest archaics to modern humans. *Time and Mind* 3:303–325.

4. Churchill, S. E., et al. 2009. Shanidar 3 Neandertal rib puncture wound and Paleolithic weaponry. *Journal of Human Evolution* 57:163–178.

5. Wynn, T., and F. L. Coolidge. 2004. The expert Neandertal mind. *Journal of Human Evolution* 46:467–487.

6. White, R. K. 2003. *Prehistoric Art: The Symbolic Journey of Humankind.* New York: Harry Abrams.

7. Tulving, E. 2002. Episodic memory: From mind to brain. *Annual Review of Psychology* 53:1–25.

8. Pinker, S., M. A. Mowak, and J. L. Lee. 2008. The logic of indirect speech. *Proceedings of the National Academy of Sciences* 105:833–838.

9. Coolidge, F. L., and T. Wynn. 2012. Cognitive prerequisites for the evolution of indirect speech. In *Oxford Handbook of Language Evolution,* eds. M. Tallerman and K. Gibson. New York: Oxford University Press.

10. Crow, T. 1997. Is schizophrenia the price *Homo sapiens sapiens* pays for language? *Schizophrenia Research* 28:127–141.

11. Gregory, N. G. 2004. *Physiology and Behaviour of Animal Suffering.* Oxford: Blackwell Science.

12. Buss, D. M., et al. 1992. Sex differences in jealousy: Evolution, physiology, and psychology. *Psychological Science* 3:251–255.

13. Coolidge, F. L., and D. Segal. 1998. Evolution of personality disorder diagnosis in the *Diagnostic and Statistical Manual of Mental Disorders. Clinical Psychology Review* 18:585–599.

Thinking Like a Neandertal

"So easy a caveman could do it"

Recently an automobile insurance company in the United States has been running a clever ad campaign featuring misunderstood cavemen. In the television spots one and sometimes two actors are heavily made up with protruding brow ridges, large faces, and unruly shocks of hair—a not-so-subtle attempt at depicting Neandertals. Otherwise, however, the actors are dressed normally and engaged in perfectly ordinary twenty-first-century activities: playing tennis, catching a flight at an air terminal, and so on. The attention-getting device is the actors' frustrated response to seeing a print ad or billboard claiming that signing up for the company's insurance is "so easy a caveman could do it." These cavemen are sophisticated, urbane young adults, not dummies, and understandably take offense at the ads. The humor works and gets our attention because it violates the folk image of what cavemen should be like. But there is actually an interesting challenge underpinning these ads, a challenge that has much to do with the content of this book: How would Neandertals fare in the modern world? And the flip side of this, how would we fare in the Neandertals' world? Both questions are in essence thought experiments about cognitive similarities and differences and provide a useful summary of our argument.

Trading Places

Introducing a Neandertal adult into the modern world would have a very different outcome from raising a Neandertal baby in a modern household. A Neandertal adult would have been born and raised in the Neandertal world, and his or her knowledge, personality, intelligence, and even physique would have been the result of that experience. An adult is a complex result of nature *and* nurture, and it is very difficult to tease apart the varying contributions of each. Thus, although it may be interesting to speculate about the immense problems a Neandertal adult would encounter, it would not make a particularly good summary for this book, whose focus has been primarily on the cognitive potential of Neandertal's genetic endowment. Instead we pose a different question: How would a Neandertal newborn fare if adopted by and raised in a modern family? By focusing on a newborn, we level the playing field and supply our Neandertal infant with all of the experiences of modern life.

We must first take a stand on our Neandertal's genetic endowment for language. The modern human's endowment consists largely of some kind of innate language acquisition device (LAD) or capacity that comes online in toddlers and guides their recognition and acquisition of the sounds used in their parents' language, the rules for making words, and the syntactical rules for stringing them together. The LAD then slowly shuts down. If a child is not exposed to a language before adolescence, as in the rare and sad cases of children raised in complete isolation, he or she will never be able to learn a language. The LAD is an innate template with a timed activation. Did Neandertals have one? This is the key question for characterizing Neandertal communication, and also a key question for our thought experiment. We are not without evidence. We know that Neandertal brains had a well-developed Broca's area and that their throat and facial anatomy was capable of speech. Culturally Neandertals demonstrated technical and hunting systems that required information exchange, and their use of pigments and the occasional pendant suggests a symbolic capacity. These positive points must be weighed against the small size of Neandertal social groups, the local nature of their way of life, the absence of any evidence for a narrative tradition (no depictions, no social use of fire), and as always, their 500,000-year evolutionary separation from us.

The conservative interpretation of this evidence is that Neandertals had an LAD, some components of which were inherited from *Homo heidelbergensis*, but that this LAD differed from ours. It could have been very different, or just a little different. We just do not know. But even granted an LAD different from ours, there is still reason to believe that our adopted Neandertal would eventually become competent in his or her parents' language. Millions of people learn a second language as adults, years after their LAD has shut down. To be sure, adults do have a harder time learning a second language, and there is a lot of variability in the level of competence attained, a variability determined largely by their personal situation. But almost all adults can do it.

Of course, our Neandertal child would not bring the same resources to the task of language learning as would a modern adult. The adult would already know one language, which would offer at least an initial framework for learning the second. Curiously, brain imaging results suggest that adult language learners often rely on different parts of the brain than the ones they use for their primary language. Our Neandertal child would also have his or her own LAD, and even if it was different from the modern LAD, it would likely be of some use. Thus even though it might take longer, we think our adoptee would eventually become competent.

When our Neandertal infant grew into a toddler he would begin to acquire the words of his parents' language (for rhetorical ease we will assume our adoptee is male), with a timing and sequence indistinguishable from that of his modern siblings. His vocabulary size would, if anything, be larger than that of his siblings, and he might be particularly adept at learning stock phrases. He could learn basic rules of syntax. Language itself acts as a structure on which children can bootstrap their communication abilities, and a Neandertal child could acquire many elements of modern syntax, even ones that had never been a component of Neandertal language. For the first several years there may be no obvious difference in the expressive capacity of our adoptee and his siblings. It would only be at the age when children start constructing longer, more complex sentences that our Neandertal might lag. Certain constructions might present problems: successive embeddings, counterfactuals, and hypotheticals. Languages in which verb forms must be held in working memory until the end of a long utterance might be a real trial. Our protagonist would be most comfortable with direct speech in a rich social context, the kind that still dominates most everyday speech in a modern human home. Could he learn to read? Almost certainly, in our opinion, at least at a basic level. Advanced technical writing, literature, and academic prose might be a challenge, but not basic informational text. In sum, our adopted Neandertal would be at only a slight disadvantage in verbal communication, but not to the degree that he could not attain a functional use of a modern language.

As a child our Neandertal would thrive in a modern family. All of his social cognitive abilities evolved in just this kind of environment of emotionally dependent small groups. Of course, most of ours have too; they are a shared legacy of earlier hominin evolution. Our Neandertal would be loving and supportive and have empathy for family members who are hurt or in distress. As a child he would have neither fewer nor more problems interacting with siblings and cousins. Indeed it is in this small-scale social context that we would see the fewest differences between our Neandertal adoptee and his modern siblings, and any differences that did emerge would likely show an advantage for the Neandertal.

This adeptness at family life would continue into adulthood. Our Neandertal would fall in love, with all the attendant emotional chaos, and would become a loving and devoted spouse. It is very unlikely that a Neandertal husband would ever sacrifice home life for work; home and family would be where he flourished. To be sure, his modern wife might accuse him of not talking enough, but not of not caring. The only possible dark corner to marrying a Neandertal would be that he might want more or less constant companionship. "His and her spaces" were not a Neandertal forte.

Outside of the context of family life, our Neandertal would have more difficulty, on average, than his modern counterparts. As a child he would be very shy of strangers, a not uncommon trait in human children, but he would be far less likely to grow out of it. He would almost certainly be terrified of going to school, to the degree that home schooling might be the only option. As an adult he would tend to be xenophobic and credulous. He would have a hard time distinguishing between strangers and enemies and would treat all strangers with suspicion and perhaps hostility. But once he determined that a stranger was in fact a friend, he would be generous and affectionate, making him susceptible to con men, telemarketers, and flimflam of all sorts. And he probably would not have cheater-detection ability or a cost-benefit ability. He would not be good with money. As much as we would like to make amusing allusions to politicians, our Neandertal would be very unlikely to be attracted to public life of any sort, and if he were, he would not be good at it. It is not his lack of complex syntax that would handicap him (at least for some constituencies), but his tendency for direct speech. He would not be very good at code switching and diplomatic speech and would be unwilling and unable to negotiate effectively with enemies.

Could our Neandertal adoptee grow up to be a successful participant in the modern industrial and information age? This is largely a question of vocation. What jobs or careers would be amenable to his abilities? Let us first suppose that our Neandertal sought the advice of a career counselor and was given an aptitude test. The result would likely show high scores for mechanical aptitude but low scores for verbal ability. His quantitative aptitude would be a mixed bag. He might score high in arithmetic calculation and intuitive geometry (e.g., those annoying unfolded box tests), but on more formal math he would probably struggle. He could memorize formulas, but knowing how to apply them would severely tax and probably overload his working memory capacity. On social cognitive measures he would score high on empathy but low on levels of intentionality. His complete profile might not look all that unusual; it certainly wouldn't appear pathological or unemployable. So where should he turn?

Rather than compile an arbitrary job-wanted ad from a Neandertal society newsletter, we will focus on a few careers in which our Neandertal would not just succeed, but perhaps excel. One that would play to all of his strengths (real and metaphorical) would be commercial fishing. Whaling, crab fishing, trawling for salmon or tuna, all still rely on the tactical hunting abilities and expert procedural knowledge that were Neandertals' forte. Our Neandertal's physical strength and ability to learn motor procedures would carry a clear advantage. And here we include not just working as a deck hand, but also acting as captain. A fishing captain must rely on an encyclopedic knowledge of local territories and conditions. Some of this is verbal knowledge, but much is not. While our Neandertal captain might not be able to explain how GPS navigation works, he could certainly use it. And the small size of most commercial crews would lie in his comfort zone socially. The only problem that he might have would be in negotiating a sale price for the catch. He would be fairly easily conned.

A second obvious career choice would be mechanical maintenance: auto repair, aircraft maintenance, and so on. We suspect that a Neandertal could learn the ins and outs of even the most complex machinery and be able to diagnose and troubleshoot problems. He would not be especially good at inventing novel, unorthodox solutions, but he could learn and apply all standard procedures. An even better choice might be skilled craft of some kind, as a machinist, blacksmith, or glassblower, for instance. These would engage his physical strength, his expert cognition, his patience and perseverance, and his ability to learn by apprenticeship.

More controversially we imagine that our Neandertal might make a very good physician. Medical diagnosis is largely a nonverbal, expert ability based on long experience and apprenticeship. It relies on complex pattern recognition and an ability to select among a large range of learned responses and solutions, just the kind of thing Neandertals were good at. And most modern physicians are not innovative; the consequences of failure (and malpractice) are just too high. Our Neandertal would be an empathetic medical practitioner, with a particularly sensitive bedside manner. But being good at something and becoming that something are different challenges. The major roadblock to this career for our Neandertal, regardless of how good he might ultimately become, would be the undergraduate premedical curriculum. He would have a very hard time learning calculus and physical chemistry; they might well be beyond him. He would have an even worse time with "soft" courses like Eighteenth-century European Literature and Modern Social Problems. Thus, as much as we believe our Neandertal could make a good medical practitioner, the modern education system would weed him out.

Many military careers would be a good fit, especially combat branches. Today, as in the past, combat training relies on developing expertise, action and response patterns so well learned that they can be selected and executed almost automatically in situations of very high danger. Neandertals were very good at this, and when we add this to their high level of pain tolerance, stoicism, and taciturn personality, we have the profile of an effective soldier. Our Neandertal might even make a very good small-unit tactician with his ability to assess situations and deploy the appropriate response. But he would not rewrite the tactical manual nor use novel responses. Nor would he be good at strategic or large-scale tactical decisions, whose information-processing requirements would tax his working memory capacity. (One cannot help thinking of the World War I generals who continued to launch frontal assaults into massed machine guns, even when it had become clear it was a bloody failure—a very Neandertal response, that: expertise overwhelming common sense.) But he would definitely be a good man in a fight, and he would be maniacally loyal to his unit.

Our intent here is not to classify careers as either Neandertal-friendly or Neandertal-unfriendly but to point out that different jobs in the modern world emphasize different sets of skills and aptitudes and that our Neandertal would be good at many of them. There are others that would not lend themselves to our Neandertal's natural aptitude. Any career based on understanding and manipulating large groups or strangers would be a poor choice. We include here almost all jobs requiring sales or entrepreneurial activities. Our Neandertal adoptee would be naturally shy and would also lack natural abilities in cheater detection and cost-benefit analysis. To be sure, these could be learned, at least to some extent, but he would still be at a disadvantage in any kind of contract negotiation. And although we know many readers would love for us to conclude that our Neandertal would make a good politician, we cannot. Good politicians rely on precisely the cognitive abilities Neandertals did not naturally possess. The natural sciences would also not be a good fit, except perhaps at the nuts-and-bolts level of running experiments or fieldwork designed by someone else. Our Neandertal would be very uncomfortable with uncertainty and the long-range contingency thinking that is at the heart of scientific research. And although a good soldier, he would make a poor policeman, where the ability to assess strangers (not enemies) is paramount.

We wish to belabor this career topic no further, but several points deserve repeating. First, an adult's aptitude results from the combined operation of genetic endowment, cultural environment, and individual experience. This would be as true for our Neandertal adoptee as for any child. Even in the absence of a modern LAD, cheater detection, and enhanced working memory, among other things, our Neandertal could acquire the skills and knowledge to succeed in the modern world. Second, only a narrow range of modern jobs and vocations would be beyond his reach. Third, whatever difference there would be should be considered a matter of cognitive style rather than a measure of intelligence. While our Neandertal would probably not be able to argue critical theory with a sociology professor (or be interested in doing so), few modern humans could either. And there is as good a chance that our sociology professor could not troubleshoot a failing fan jet engine on a commercial aircraft to save his or her life (and ours). But our Neandertal could.

For many modern people religion is the primary institution outside the family governing their self-identity and worldview. Nothing in the archaeological record suggests that Neandertals had formal religion, or even concepts of supernatural agency or causation, so how would our Neandertal adoptee participate in modern religious life? Would he be a natural atheist? We don't think so. In fact we think that our Neandertal's participation in his parents' religious activities would be little different from that of his siblings. Much religious education is rote learning of procedures and texts, and our Neandertal would be good at this—Sister Mary Esther would approve. Children, indeed most adults, do not grapple with the theoretical mysteries of their faith; they simply learn them as true. Our Neandertal would have no difficulty here, and might well grow into a committed, devout believer. He would find great comfort in the formal rituals of some religions and would rapidly lose interest in serious metaphysical discussions. Because of his natural shyness and taciturn disposition, he would be uncomfortable in public declarations of belief, even if strongly held, and perhaps also large-group celebrations. His natural tendency toward xenophobia and aversion to critical thinking might make him a prime candidate for fanaticism. Some religions would be a better fit than others, especially religions that have identified a single truth and are not open to doubt. However, as with modern humans, our Neandertal is likely to be most comfortable with the religion of his parents.

In sum, a Neandertal baby raised in a modern human household would grow into a fully functioning human adult. He would not need institutionalization, or even need to be treated as special. His genetic cognitive endowment would be rather different from average, and this would tend to channel him in particular developmental directions, but he could acquire all of the necessary abilities for modern life. But fair is fair. How would a modern human fare in the Neandertal world?

Turnabout

It is especially revealing to run our thought experiment in the other direction. A modern adult, snatched by aliens off the streets of Paris and flung naked back in time 50,000 years would undoubtedly face grim odds. First, if Neandertals found him before he died, they would probably treat him as an enemy, that is, kill him. But even if the Neandertals' empathy led them to take in our reluctant Doctor Who, he would face a daunting challenge. Without the know-how acquired over years of growing up as a Neandertal, he would have a hard time keeping up, literally and metaphorically. Physically he just might not be able to do it, even if fit by modern standards. If he was strong enough, he would still be clueless, and his lack of a Neandertal language acquisition device, and adult age, would make it very difficult to learn their speech. For several years at least he would be a handicap to his Neandertal foraging group, and though they might become fond of him as a kind of semi-Neandertal pet, they might not be able to support him for as long as it would take to get him up to speed. By that time the cold and physical stress would undoubtedly have taken its final toll. Perhaps he would be lovingly placed in a low spot in a cave, covered over, and soon forgotten.

As with the Neandertal example, a human infant would probably fare better. The rigors of Neandertal infancy would be harder on our modern infant than on his Neandertal siblings, who would be physiologically and anatomically more suited to the stress. Modern infants have evolved for thousands of years in a context of cultural and technological buffers unknown to Neandertals, and even though our Neandertal foster mother would do her best, it would initially be touch and go for our protagonist. If he survived the first year of life, he could begin to learn how to be a Neandertal. As with our Neandertal adoptee, language learning might be a stumbling block. Our modern toddler would not have a Neandertal LAD, and even if he shared some basic features of the LAD inherited from *Homo heidelbergensis* (lexicon development, perhaps), his own LAD would lead him astray in acquiring Neandertal grammatical rules. He would be slow in learning language, by Neandertal standards, but he would eventually become proficient, just as most adult second-language learners eventually become proficient.

Education would challenge his abilities. Like all hunter-gatherers, Neandertal children learned by observing their parents and by practicing. But it appears that Neandertal children also accompanied adults in the dangerous business of large mammal hunting. They not only had to learn by observing their parents, they may also have participated at a young age. Expertise takes years and countless repetitions to acquire, and Neandertals started early. It is possible that our modern adoptee would be slower to acquire the large body of skills and procedures necessary for Neandertal technology and hunting, but he could certainly do it.

Social learning would be easier. Neandertals relied on the same small-scale, embodied social cognition that modern humans use, and our modern child would be comfortable in his Neandertal group. In the absence of other modern social models, it is unlikely that our child would develop his natural aptitude in social manipulation. His cost-benefit and cheater-detection abilities would have no context in which to develop, and while it is amusing to envision a Palaeolithic *Invention of Lying* scenario in which our precocious Machiavelli became a Neandertal flimflam artist, this would be unlikely. Our modern social abilities appear to have evolved to deal with large groups, including acquaintances and strangers, and Neandertal life just did not have these to offer.

Neandertal life offered only one career choice: large mammal hunter-gatherer. Our modern human could certainly have learned how to hunt mammoths and rhinos or cattle and gazelles, but he might never have become quite as adept as his Neandertal siblings. For one, he might not have mastered as many tactical routines and might have been a tad slower at assessing a situation and executing the optimal solution. It might not have been quite as automatic for him, and hesitation would eventually be costly. Our modern Neandertal wannabe might not survive the kinds of traumas Neandertals all eventually received. On the other hand, he might see alternatives never considered by his Neandertal friends. His enhanced working memory would naturally allow him to compare various procedures, select optimal bits from each, and come up with a new tactic. Of course, his Neandertal cousins' dogged adherence to learned routines would make them highly resistant to new ways of doing things, and his too-clever-by-half ideas would probably go nowhere.

How would our modern adoptee have fared as an adolescent in the Neandertal mating game? Much would hinge on whether or not his Neandertal territorial community was male or female philopatric. If our modern adolescent was a boy who had to leave his natal community and attach himself to another, his chances might be poor. His very odd appearance would inspire even greater hostility than normal in the men of his target community. And the local available women would be unimpressed, as well they should be. His puny build and poor fighting skills would make him a poor choice as a mate. Our adoptee might succeed in the mating game if his community was male philopatric, and eligible females moved in from neighboring communities. He would still be unappealing, but his family might present him as the only acceptable candidate. He could father Neandertal children, but they too would likely have a less than optimal phenotype. Eventually his modern genes would probably go extinct.

In sum, our adopted modern infant could, with luck, grow up to think like a Neandertal. But he wouldn't make a particularly successful one. He would not lead Neandertals out of the evolutionary cul-de-sac they had entered. His unique cognitive abilities would yield no direct advantage in the Neandertal world. They might have led him to feel strangely unfulfilled socially, and more than a bit absentminded or even downright odd, but in the absence of a modern cultural milieu, they would have provided no mental advantage.

Neandertals' Demise

There are no Neandertals living today; the last one died sometime after 30,000 years ago in what is today Spain or Portugal. Most populations, indeed most species, eventually disappear, either through extinction or through evolutionary transformation into something different enough to earn a new name. So Neandertals' disappearance was not an unusual evolutionary event. But it holds an unusual fascination for us because it was tied directly or indirectly to the actions of our own ancestors, *Homo sapiens sapiens*.

The evolutionary story of modern humans is now well understood, at least in its major outlines. We evolved in Africa out of a local population of *Homo heidelbergensis*. By 200,000 years ago people with modern anatomy were living in Ethiopia, and by 100,000 years ago these people began to leave traces of modern-like behavior. Sometime after 70,000 years ago our ancestors expanded out of Africa, and in the evolutionary eye-blink of 60,000 years had ventured to all corners of the habitable world, including North and South America and Australia. Along the way they encountered and supplanted many local populations of archaic humans. Among these were Neandertals living in Europe and western Asia. This story of modern humans is often presented as a tale of evolutionary success, of the triumph of brains over brawn. Here we would like to present it as a sad tale of cultural incompatibility and evolutionary doom. We take as our perspective that of a Neandertal community living in central Europe 40,000 years ago.

The Fearsome Others

The first hint of something new and troubling probably came to our Neandertal community with the arrival of an adolescent girl from a neighboring community to the east, looking to find a husband and start her adult life. She carried a troubling account of dark strangers seen to the east of her natal home. Perhaps she had not herself seen them but had heard about them from her brother's wife. Neandertals were not really given to idle chitchat or storytelling, so the rumor probably died. At worst it created a vague anxiety in the naturally xenophobic Neandertals; it did not inspire a desire to find out more. Perhaps the rumor was remembered again years later, with the arrival of another adolescent, but this time one who had herself seen them. What account might she have given, from her perspective as a Neandertal woman?

The foreigners, whom we will dub Cro-Magnons, had a very strange appearance. They were tall, several inches taller than the tallest Neandertal man. They had dark skin and dark hair, very unlike the light hair and eyes of Neandertals. They covered themselves head to toe in tight-fitting, tailored clothes. When they removed their clothes their bodies were thin and poorly muscled. (Our Neandertal girl would probably have laughed.) But their heads were oddest of all. They had bulbous heads with very small, child-like faces, small noses, and ugly protruding chins.

At first our Neandertals would have seen only groups of men hunting for animals, men who used hunting techniques that were unlike those of the Neandertals. Instead of relying on large animals who frequented a small region, these Cro-Magnons often journeyed long distances to intercept game at distant spots (and in so doing, crossed through several Neandertal territories). And they attacked animals from a distance, propelling their spears with a device that extended their arms. The spears themselves were long and flimsy, with projectile points made of bone or antler instead of stone. These took many hours to produce, instead of the few minutes required for a Levallois point. More surprising was the absence of women and children in these hunting parties. Where were they, and what were they doing? Eventually, as these Cro-Magnons moved closer, our Neandertals would have seen parties of women and children foraging apart from the men. They didn't try to kill large animals, but instead focused on small game, often using long nets to capture small mammals. They even occasionally built devices that killed or captured animals when the people were not even nearby; the women would return the next day to collect the results.

Eventually these Cro-Magnons were no longer a distant rumor, but had moved right into the regions frequented by our Neandertals. With more opportunities to observe, the more audacious Neandertals were able to see even more oddities. Cro-Magnons' home life was very different from Neandertals'. Their face-to-face groups were much bigger, not just a few related families, but scores of people living together. They sometimes lived in caves that Neandertals had used, but in these cases the Cro-Magnons spread out along the cliff face on both sides of the opening. They also often built large fires, and older men and women kept the fires burning through the day while others foraged. Groups of people sat around these fires long after cooking was done. Sometimes one person would talk for hours, holding the others' rapt attention. And sometimes they would dance, using drums to keep rhythm and bird bones to make whistling sounds.

The Cro-Magnons wasted much of their time decorating themselves and making useless items. They put small pieces of polished ivory on strings around their neck and even through their ears. Strangest of all they spent hours making small versions of animals out of ivory and bone. From our Neandertals' perspective some of this activity would be puzzling, but some would also be intriguing or attractive. Neandertals did alter their own appearance with pigments, and they might well have taken to the idea of using beads or pendants to enhance their personal attractiveness. Had a Neandertal found a lost figurine, he or she might well have kept it as a curio, much as pretty crystals were collected. But a Neandertal would see little point in duplicating the production. Neandertals did not have a rich symbolic culture or a narrative tradition, and their children did not play with toys. There would be no role for the figurine to play. Similarly Neandertals would not have produced cave paintings; they did not venture deep into caves, and had no symbolic, religious life to celebrate or actuate.

Our Neandertals would be puzzled by the newcomers' interaction with strangers. The Cro-Magnons often traveled outside their home territories to visit people in distant places. They carried gifts with them, and returned days and weeks later with artifacts and materials acquired in distant territories. And strangers visited in return and were not treated with hostility. Most disturbing of all, the Cro-Magnons tried to interact with our Neandertals. At first the Cro-Magnons tried a direct approach, using open-handed and open-armed gestures of peaceful intent. Our Neandertals fled or made hostile gestures in return. The Cro-Magnons may have tried a delayed exchange, leaving goods in the open for Neandertals to take. The Neandertals took them, but left nothing in return; economic negotiation was not a component of their cognitive tool kit. Eventually the Cro-Magnons gave up the attempts.

One of the oddest facts of Neandertal and Cro-Magnon cultural history is that the two groups shared western Europe for several thousand years, and yet there is very little credible evidence for interaction. The little gene flow that occurred between them had happened 20,000 years earlier in the Middle East, when they had first come into contact, and it appears to have been one-way, Neandertal to modern; archaic Neandertal DNA has no modern human DNA. Neandertal technology did not borrow anything from the Cro-Magnons. It is not that the Neandertals could not have made such things if they chose to; they just didn't. But Cro-Magnons didn't borrow much, if anything, from Neandertals either, which is odd. Modern humans have always been good at borrowing useful knowledge from indigenous populations when they came into contact. Surely the Neandertals had local knowledge that would have been of immediate use to the Cro-Magnons. Yet we see no evidence for such borrowing. It appears as if Neandertals and Cro-Magnons had a tacit agreement to avoid one another entirely. Indeed there are reasons for thinking Neandertals and Cro-Magnons may have been different enough in their ways of life that they could avoid one another and both be successful—for a time.

Earth's climate began to cool dramatically about 30,000 years ago, and in Europe the northern glacial ice began to push south again. As it got colder the animals moved. Neandertals had coped with this kind of thing before and shifted their focus to new kinds of animals as conditions changed. But this time they were not the only human predator. Cro-Magnons also had to change, and they did, but much faster and much more dramatically. Over the course of a few years, not the few generations typical of Neandertals, they devised new strategies and new technologies. And whereas before Neandertals and Cro-Magnons could live in balance, now they no longer could. Exploitation of the same resources put them into direct competition. There is no evidence for violent encounters, but these must have occurred. A Cro-Magnon man would be no match for a Neandertal in a fair fight, but fights were probably few and rarely fair. Cro-Magnon hunting parties were too big, and they had missile technology. Neandertals were forced to retreat. Eventually they found themselves cornered on the Iberian Peninsula, where they held out for a while, but their way of thinking was just unable to respond ecological move for ecological move with the Cro-Magnons'. Perhaps the Neandertal adaptation was too committed to well-learned routines, too focused on specific targets on the landscape. In direct competition with Cro-Magnons they needed to shift focus to new foods and invent new techniques. Neither ability was a well-developed component of Neandertal cognition. Or their inability to respond may have had nothing to do with cognition. Their territorial communities were small and scattered, and when a territorial community became isolated from others, with no access to mates, it soon perished. After a few thousand years of this, the Neandertal population was reduced beyond the point of possible demographic recovery. They dwindled away and were lost to the world.

Neandertals live on today only in our dimmest cultural memories. It is just possible that European folk traditions of trolls, Cyclops, and even dwarfs have roots in the ancient encounter between Neandertals and Cro-Magnons. But tales of strange human-like creatures are common the world over, and Neandertals were not. Today Neandertals are star players in a science-based narrative. When French Neandertals were described by paleontologists over a century ago, they created a public sensation, and Neandertals quickly entered general cultural knowledge through fictional accounts in novels and film. They inhabit this world still. Thus in stereotyped form they live on.

Our students often ask us whether science will ever clone a Neandertal, as if this would inherently be a good thing. They ask because they are fascinated by the existence of humans who led very different lives but who were still, somehow, like us. This is the ultimate fate of Neandertals: to live on as inexact mirrors of ourselves.

Glossary

Abri Pataud A rockshelter in southwestern France dating to about 23,000 years ago, where a rock-lined hearth was found that has been associated with *Homo sapiens*.

Abric Romani A rockshelter located in northeastern Spain with evidence of Neandertal hearths that date to between 40,000 and 70,000 years ago.

Acclimatization Non-genetic, developmental adjustments of the body to local conditions such as temperature and altitude.

Allen's rule In evolutionary science, the tendency for mammals living in cold habitats to have relatively short *limbs* and mammals in hot habitats to have relatively long limbs. These proportions result from natural selection optimizing limb length for heat retention or cooling.

Allocentric Characterized by perception of an object or scene from a perspective or point of view other than one's own.

Antisocial personality disorder Also called psychopathic personality (or psychopath) or sociopathic personality (or sociopath). These people have little or no regard for the feelings or lives of other people, even children, and thus they are very dangerous.

Appropriate incongruities The idea that we naturally think that something is funny because it is absurd or nonsensical in some way, like "...a kangaroo walked into a bar..."

Artless Dodger A play on the name *Artful Dodger*, a character in a Charles Dickens' novel, who was a clever pickpocket. Thus, *artless dodger* is a pun implying that someone is clever but does not produce works of art.

Asperger's disorder A type of autistic disorder but without language delays in childhood.

Atlatl A tool that extends the arm, and is used to throw spears greater distances than is possible with the arm alone.

Australopithecus afarensis A species of extinct hominins that lived in Africa about 3.5 million years ago and which is possibly ancestral to *Homo erectus*. Its upper body had features similar to those of apes, but *Australopithecus afarensis* walked on two feet. The most famous specimen (skeleton) of this species is nicknamed *Lucy*.

Autism A disorder of the development of the brain that is characterized by the inability to interact socially and emotionally with other people, communicate with other people, and by limitations in and repetitions of behavior.

Autobiographical memory The information that the mind holds that consists of things like events that happened in past experience; also known as episodic memory.

Autonoetic thought Also known as *autonoesis*. It means an awareness of the arbitrariness of time, in that people can travel in their minds in time by going to memories of past events or imagine future events.

Bergmann's rule In evolutionary science, the tendency for mammals living in cold habitats to have compact bodies and mammals in hot habitats to have relatively elongated bodies. These proportions result from natural selection optimizing body build for heat retention or cooling.

Bipedal Walking on two feet.

Blombos Cave A 77,000-year-old site on the southern coast of South Africa that has yielded some of the earliest evidence of personal ornaments, shell bead necklaces, as well as engraved objects and bone tools.

Bonobo *Pan paniscus*, a species of chimpanzee that is smaller than the Common Chimpanzee (*Pan troglodytes*) and which differs in some important behavioral characteristics as well.

Brain waves Varying kinds of electrical activity that are produced by the brain and which can be measured through electrodes placed on the scalp.

Broca's area A front area of the hominin brain linked to the production of speech and other functions.

Bush baby A species of small (up to 12 ounces), nocturnal primate (prosimians) that is native to Africa.

Cardinal directions The four orientations known as north, south, east, and west.

Carnivore An animal whose diet consists of the meat of other animals.

cc Cubic centimeter. A golf ball has a volume of about 41 cc, a hard baseball about 212 cc, a regulation softball about 478 cc, modern *Homo sapiens sapiens* brains about 1350 cc, Neandertal brains about 1500 cc, and a basketball about 7284 cc.

Cheater detection The ability to tell when another person is acting dishonestly (lying, deceiving, manipulating, or being unfaithful).

Chimpanzees A species of Great Ape that lives in West, Central and East Africa, and which is humankind's closest living relative. Chimpanzees and modern humans are thought to have a common ancestor that lived about 8 million years ago.

Code switching The practice of moving between different versions of a language (for example, between formal and ethnic English, as in *hello* and *yo*) or between two different languages (for example, between English and Spanish, as in *hello* and *hola*) in different contexts.

Contra coup effect A blow to one side of the head propels the brain inside the skull to hit the opposite side of the skull (away from the injury site) and, many times, the *contra coup* site exhibits greater dysfunction than the actual site where the skull was initially injured.

Counterfactual A what-if sentence that stands against the facts; an example is, "If I were king, I would pay you a million dollars."

Cranium The part of the skull that surrounds the brain.

Cro-Magnon An informal term referring to modern humans (*Homo sapiens sapiens*) who entered Europe about 40,000 years ago. The name comes from skeletal remains found in 1868 in the Cro-Magnon rock shelter at Les Eyzies, France.

Declarative memory A memory for words and their meaning. Two types of declarative memories are commonly listed: *semantic memory*, for the recall of facts and lists of words, and *episodic memory*, for the recall of personal experiences, often recalled like a story with time, place, and feeling characteristics.

Diplomatic speech Also known as *indirect speech* or *polite speech*. It is speech that hides the speaker's real meaning or real feelings, but still gets the speaker what he or she wants from another person.

Disorder A term in psychology that means the behavior is harmful to one's self or other people. The term *disorder* also means that the person cannot hold a job, stay in school, or get along with other people.

DNA Deoxyribonucleic acid, an acid that contains the genetic blueprints for all living organisms (except for some viruses).

Dolni Vestonice A *Homo sapiens sapiens* living site in the Czech Republic dating to between 27 and 24,000 years ago where archaeologists have excavated extensive evidence for daily life, including evidence for ritual activities.

Dominant gene A single gene that influences the physical characteristics of offspring.

Doctor Who A fictional time traveler in a long-running BBC science fiction show. A "reluctant Doctor Who" is a time traveler who is not happy about being kidnapped and flung backward in time.

Embodied cognition The idea that the physical body and all its senses are part of how we think, and that tools and other utensils (like a blind person's walking stick) also influence our perceptions of the world.

Endocast The internal cast of a fossilized skull, which may show some features and shape of the brain that was once inside the skull.

Episodic memory The information that the mind holds that consists of things like events that happened in past experience; also known as autobiographical memory.

Evolutionary psychology A branch of psychology that focuses on the role of natural selection in the evolution of the brain, mind, and behavior.

Executive functions of the frontal lobes A collection of abilities of the front part of the brain that include paying attention, actively inhibiting bad behavior, making decisions, planning, putting things in order, and other abilities.

Foraging The act of looking or searching for food or provisions, as opposed to growing crops, farming, or domesticating animals.

Fossilization The process whereby minerals in the soil replace the chemicals in the bones (or sometimes the flesh) of dead organisms, and it can take tens of thousands of years. Thus, many Neandertal skeletons are only partially fossilized, so there may still be some organic material (flesh, skin, and non-fossilized bones) in their remains.

FOXP2 A gene in humans, mammals, birds, and other species that has been identified as contributing to the ability to communicate. It has also been found in Neandertal DNA. Mutations in FOXP2 cause severe speech problems in modern humans and communication problems in songbirds.

Franco-Cantabrian tradition The oldest of several artistic traditions in northern Spain and southwestern France that dates to about 34,000–14,000 years ago. It includes the spectacular cave paintings of animals and other images in Chauvet and Lascaux, France and Altamira, Spain.

Gene The unit of heredity for all living organisms. Genes are located on an organism's DNA.

Gesher Benot Ya'acov A site in northern Israel that boasts the earliest generally accepted evidence for the intentional use of fire, about 790,000 years old.

Ghengis Khan He was a powerful central Asian warlord who lived in the late 1100s. He founded and ruled the Mongol Empire, the largest empire in history.

Go An ancient Chinese board game played with small black and white stones.

Gracile Characterized by a delicate or lightly built anatomy, usually in comparison to a more heavily built related (robust) form (e.g., modern humans vs. Neandertals).

Hafting Affixing one tool to another, as in joining a spear point or axe head to a shaft.

Herbivore An animal whose diet consists of plants such as grasses and leaves.

Herto A site in north-central Ethiopia that dates to about 160,000 years ago. The remains of three skulls with many anatomically modern human features were found there; these skulls appear to have had the flesh removed from them, after which they may have been carried around in bags.

Höhlenstein-Stadel figurine A lion-headed, human-like statue carved out of ivory that dates to about 32,000 years old. It is the oldest known sculpture that combines elements of a human with elements of an animal. It was found in 1939 in a cave in southwestern Germany.

Hominid A shorthand term for a member of the family *Hominidae*. In Linnaean taxonomy it includes humans and fossil ancestors and relatives of humans. In modern cladistic taxonomy it also includes all of the great apes. Some paleoanthropologists continue to use it in its Linnaean meaning, but many others have switched to using hominin.

Hominin A shorthand term for a member of the tribe *Hominini*. In modern cladistic taxonomy it includes humans, fossil ancestors and relatives of humans.

Homo erectus Literally, *upright man*. A species of fully bipedal hominin that first appeared in the fossil record about 1.8 million years ago in Africa. It was considerably larger than earlier hominins, had a cranial capacity ranging from 800 cc to 1100 cc, and had an adaptive niche that allowed it to invade many different habitats in Africa, Asia and Europe.

Homo heidelbergensis Probable common ancestor of Neandertals and modern humans that lived from about 600,000 years ago to about 250,000 years ago in Europe and

Africa. It had a cranial capacity of about 1100 cc to 1300 cc, thus overlapping with the range of modern humans (1000 cc to 2000 cc).

Homo neanderthalensis See Neandertal.

Homo sapiens Literally, *man* [human] *who knows*; the species that is us, all modern people. Our species is also known as *Homo sapiens sapiens*. The first *sapiens* is the species, and the second *sapiens* is the subspecies. Thus, sometimes Neandertals are referred to as *Homo sapiens neandertalensis*, indicating that we and Neandertals are the same species but not the same subspecies.

Hypnagogic state A dream-like state between being awake and falling asleep that occurs when a person starts to fall asleep.

Hypnic jerk Also called *hypnagogic jerk*. An involuntary twitch of the muscles that usually occurs when a person starts to fall asleep.

Hypoplasia Incomplete development of a tissue, organ, or teeth. It can be caused by poor childhood nutrition, or it can be inherited genetically.

Hypothetical A what-if sentence, for example, "What if we went this way instead of that way?"

Iberian Peninsula The land that is now Spain and Portugal.

Icon A type of sign that conveys its meaning by *looking like* the object that it refers to. An example is found in the roadside signs showing the figure of a deer to convey the meaning 'deer crossing'.

Index A type of sign that conveys its meaning by some *association with* object that it refers to. An example is found in a hoof print, which points to the recent presence of the deer that made it.

Inhibition The ability to stop bad behavior or behavior that might be judged harshly by others.

Jabberwocky A poem of nonsense words by Lewis Carroll, author of *Alice in Wonderland*.

Kali The Hindu goddess of time and change, and she is usually associated with images and themes of death.

Kanzi A male bonobo (a type of chimpanzee) who learned to communicate with people using signs, and also learned to knap stone.

Kebara Cave A cave on Mount Carmel in northern Israel occupied about 50,000 years ago by Neandertals. The site has yielded the remains of 23 Neandertals, including two burials, as well stone tools, and remains of cooked food.

Krapina A cave in Krapina, Croatia where Neandertals bones showing evidence of cannibalism were found.

La Chapelle aux Saints A site in south-central France where one of the most famous Neandertal skeletons was discovered; he is also known as the "Old Man of La Chapelle aux Saints."

La Cotte de St. Brelade A site on the Isle of Jersey in the English Channel where Neandertals drove mammoth and rhinoceros over a cliff and then butchered them.

La Ferrassie A cave in La Ferrassie, France, where the remains of several Neandertals were discovered in 1909. The site also includes evidence of Neandertal corpse protection activities.

Language A system of communication based on the use of symbols and rules for stringing them together (grammar).

Language acquisition device (LAD) The natural mental capacity that allows a human infant to acquire and produce language easily.

Lazaret Also known as Grotte du Lazaret. A prehistoric cave site in southeastern France near the Mediterranean coast that has the remains of a Neandertal hut built about 160,000 years ago.

Levallois In lithic technology, a technique in which the stone knapper prepares a core in order to control the size, shape, and thinness of the flakes that are removed.

Lexigram A visual symbol representing a word that has been used to develop communication with bonobos and chimpanzees.

Long-term memory The information that the mind holds that can be recalled for longer than a few minutes and much longer periods of time, like years.

Lucy The name given to the skeleton of a female *Australopithecus afarensis* found in east-central Ethiopia. She lived about 3.2 million years ago.

Machiavelli An Italian philosopher, politician, and writer during the Renaissance whose name has become linked with being sneaky, conniving, and greedy. A "precocious Machiavelli" is someone at a young age who is especially good at being sneaky, conniving, and greedy.

Marjorie's core A Levallois core from Masstricht-Belvédère, a site in southeastern Netherlands, that has been reassembled from its recovered flakes. It was named after the first person who put it back together.

Masstricht-Belvédère A site in southeastern Netherlands, where Marjorie's core was found and reassembled.

Mitochondria A membrane-enclosed organelle found in the cytoplasm of a cell that generates most of the cell's energy.

Mitochondrial DNA (mtDNA) Double-stranded DNA (genetic material) that resides in the mitochondria of a cell. As most other DNA is found in the cell nucleus, it thought that mtDNA has a separate evolutionary origin than nuclear DNA. In mammals and most other organisms, mtDNA is inherited only from the mother. In DNA studies of recent human evolution using either mtDNA or Y-chromosomes, the former can only trace maternal inheritance patterns and the latter only paternal patterns. Interestingly, they often give slightly discrepant evolutionary time estimates.

Molodova 1 A site above the northwest corner of the Black Sea where Neandertals made stone tools and windbreaks dating back to about 45,000 years old.

Monkeys Anthropoid Primates that have tails. Monkeys and modern humans are thought to have a common ancestor about 30 million years ago.

Moula-Guercy A cave site in southeastern France where evidence of Neandertal cannibalism has been found that dates to about 100,000 years ago.

Muscle atonia The natural paralysis of the body that occurs in REM sleep, preventing the person from acting out movements in response to dreams.

Neandertal Also known as *Homo neanderthalensis* or *Homo sapiens neanderthalensis* (can also be spelled Neanderthal). A robust human type that originated in Europe about 250,000 years ago and persisted until after 30,000 years ago. Their average cranial capacity was 1427 cc with a range of 1250 cc to 1700 cc, the largest of any *Homo*.

Neophobia Also known as *misoneism*, the term used by psychologist Carl Jung. It means a fear of doing things in new ways.

Neurotic A term in psychology that means a person is anxious and tends to distort and misinterpret reality but not as badly as a psychotic person would.

New World monkeys Monkeys that are native to Central and South America, such as marmosets, tamarins, and capuchin, squirrel, howler, and spider monkeys.

Obsessive-compulsive disorder A term in psychology that means a person has thoughts that won't go away that lead him or her to be anxious and causes them to repeat meaningless behaviors, like washing their hands hundreds of times a day and checking on things over and over.

Old Man of La Chapelle aux Saints One of the most famous Neandertals ever discovered; he suffered from a large variety of injuries and ailments, including arthritis, degenerative joints, broken bones, and tooth loss. He is estimated to have lived about 60,000 years ago.

Old World monkeys Monkeys that are native to Africa and Asia, such as baboons, macaques, vervets, and langurs.

Omnivore An animal that eats both plants and meat.

Ontogeny The development of any organism from its start as a fertilized egg to its fully adult form. An old saying in biology was that "ontogeny recapitulates phylogeny," which means that watching the development of an individual organism provides insight into the evolution of its species. It is possible that sometimes ontogeny does recapitulate phylogeny.

Orkney Also known as the *Orkney Islands*, a chain of small islands located off the northern coast of Scotland that have been inhabited for about six thousand of years.

Ortvale Klde A rockshelter in the Caucasus mountains in Georgia (extreme southeastern Europe) where Neandertals hunted a local species of mountain goat, the Caucasian tur, about 50,000 years ago.

Paleoanthropology The study of ancient humans through fossils, faunal remains, and stone tools, with a particular emphasis on humans and hominins who lived over 20,000 years ago.

Paleo-Indian Term for the earliest Americans, thought to have migrated from Asia sometime around 15,000 years ago. They were expert big-game hunters who used a sophisticated system of spear throwers and spears with detachable foreshafts.

Peninj Site in Tanzania, Africa, where about 1.4 million years ago, ancient hominins used stone tools to scrape wood.

Perikymata Microscopic growth lines that appear on the surface enamel of teeth as a series of groves; they are used to assess the age of the individual who grew the tooth.

Personality disorder A term in psychology that means a person mostly annoys, bothers, or frightens other people but tends not to be bothered by his or her own behavior.

Philopatry Remaining in or returning to an individual's birthplace. In *male philopatry*, individual males remain in their father's birthplace to mate and produce offspring, where in *female philopatry*, individual females remain in their mother's birthplace to mate and produce offspring.

Phylogeny The development and change of any species through its evolutionary history.

Pinnacle Point A cave site in southern South Africa that shows evidence of occupation between 170,000 and 40,000 years ago, where beads and stone tools have been found dating to about 100,000 years ago.

Pirahã A tribe of hunter-gatherers from the Amazon region of South America. They speak a language (also named Pirahã) that is unusual because it appears to lack words for numbers and colors, as well as other features commonly associated with language.

Procedural memory The information that the mind holds that consists of knowing how to do things, such as juggling, riding a bicycle, or stone knapping.

Prosimian Prosimians are primates that are not monkeys or apes, and they are more evolutionarily ancient than monkeys and apes. They include, among others, lemurs, bushbabies, and tarsiers.

Protolanguage A language that is ancestral to and possibly simpler than modern languages.

Psychopathic personality (or psychopath) See antisocial personality disorder.

Psychotic A term in psychology that means being out of touch with reality.

Pyrotechnology The intentional use and control of fire by humans.

Recessive gene A single gene that does not influence the physical characteristics of off-spring unless it is paired with a similar gene from the other parent. If it is unpaired, a recessive gene will remain hidden in the offspring's DNA, where it has the potential for influencing future generations of offspring (if paired with a similar gene from a future parent).

Recursion In language, the feature that inserts a word or phrase inside another word or phrase, that then modifies the meaning of the entire word or phrase.

REM Rapid Eye Movement, a stage of sleep (25% of human sleep) in which dreams are vividly experienced.

Retromolar space The space at the back of the jawbone (mandible) between the last tooth (molar) and the ascending part of the jaw. It is not always present in Neandertals but helps to identify a specimen as Neandertal when found.

Ritual A set of actions that are performed in a rigid, invariant manner in specific circumstances. They often have symbolic value. Rituals can include such practices as treating and disposing of (such as burying) the corpses of people who have died.

Robust Characterized by a heavily built anatomy, with thicker bones and evidence of greater muscularity; usually in comparison to a more lightly built related (gracile) form (e.g., Neandertals compared to modern humans).

Salzgitter-Lebenstedt A site in northern Germany where Neandertals hunted and butchered raindeer.

Schizophrenia A mental disorder characterized by delusions (false beliefs) and hallucinations (hearing or seeing things that aren't really there).

Schöningen A site in north-central Germany where several ancient wooden spears were found that date to about 400,000 years ago.

Semantic memory The information that the mind holds that consists of names, words, facts, and so on.

Sexual dimorphism The difference in the body size and features of non-reproductive anatomy between males and females of the same species. In primates, greater sexual dimorphism (males being much larger than females) appears to correlate with greater aggression between males.

Shaman A religious practitioner who relies on personally acquired supernatural power; also known as medicine men or medicine women. They are often thought to be messengers between the human and spirit worlds, who often also treated illnesses and injuries.

Shanidar A cave in the Zagros mountains of northern Iraq where at least nine Neandertal skeletons have been found since 1953; they date back to about 50,000 years ago. The Neandertal skeletons found in this cave are each named after it and are numbered (i.e., Shanidar #1, etc.).

Short-term memory The information that the mind holds for only a few seconds but usually less than a few minutes.

Sign Something that stands for something else. Signs include words, images, gestures, smells, tastes, textures, and sounds. The cognitive abilities required to extract meaning from the various types of signs (icons, indexes, and symbols) are different for different animals (including humans).

Sign language Any language that uses finger and arm gestures rather than sounds to communicate words and ideas.

Sima de los Huesos Literally, *pit of bones*. A site in northern Spain where the bones of more than thirty *Homo heidelbergensis* have been found that date to about 400,000 years ago; the individuals appear to have fallen or have been dropped down a deep shaft in the cave.

Social cognition A term in psychology that refers to any thinking processes (reasoning, memory, etc.) that we use to interact with people.

Sociopathic personality (or sociopath) See antisocial personality disorder.

St. Cesaire A rockshelter in the Charente-Maritimes department of southwestern France where 35,000-year-old Neandertal skulls were discovered in 1979.

Stoicism Enduring pain or hardship without showing feelings or complaining. Named after a school of early Greek philosophers, who did not like to show their feelings and did not like to complain about pain or hardships.

Stone knapping In stone tools, it is the process of removing stone flakes from a stone core by hitting the core or pressuring a flake off of the core.

Sungir A burial site in Russia where three modern humans, an adult and two juveniles, that date to about 27,000 years ago were found, along with thousands of beads and ornamental ivory disks and spears.

Symbol A type of sign that neither looks like the object that it refers to nor points to it in any way; instead, the meaning of a symbol comes from a *social agreement* (as in language). An example is the word 'deer' (an arbitrary sound) standing for the animal 'deer' (the object designated by the sound through an agreement in language). Symbols may also be highly personal and may only have some meaning for the maker of the symbol.

Syntax The rules in language for making up sentences; an example is the rule in the English language that subjects of sentences generally come before the verb.

Tata A site in northwestern Hungary where an engraved pebble was found that has been argued to have been purposefully engraved by a Neandertal (a simple line crossing a natural groove).

Theory of Mind (ToM) The ability to understand another person's thoughts, attitudes, or beliefs.

Tur A species of mountain goat that lived in the Caucasus mountains of Georgia (Russia) and was hunted by Neandertals about 50,000 years ago.

Ungulate Any kind of animal with hooves, like reindeer, deer, and goats, that were a primary choice of Neandertals for hunting.

Werner von Braun A German-American rocket scientist who was one of the leading pioneers in developing the American space program.

Working memory There are two different meanings of the term working memory. One is a kind of short-term verbal memory, stored and then recalled only briefly, like a phone number. The other kind is a complex model of memory with four components (a central executive [which helps you to pay attention to things], a phonological store [which stores what you hear], a visuospatial store [which stores what you see and directions], and a memory store [that stores and integrates the content of the other components]). In this sense, working memory is information that can be held in mind and manipulated, despite interference (distractions).

Index